# The Phylogeography of Afro-Americans and Africans

by

CLYDE WINTERS

# DEDICATION

This book is dedicated to the many researchers who have asked me to write a book about the Phylogeography of African people.

# Table of Contents

Dr. Clyde Winters

## Abbreviations

**kya** thousand years ago

**TMRCA** the most recent common ancestor

**amh** anatomically modern human

**BP** before the present

**OoA** out of Africa

**Hg(hg)** haplogroup

# Introduction

In this book we will discuss and explore the genetic ancestry of African and Afro-Americans. You may upon reading this, say why am I differenciating between Afro-Americans and Africans. The answer is simple, Afro-Americans are not just the descendants of African slaves, their ancestors also include Black Native Americans and Black Europeans who came to America from Europe as indentured slaves and freemen.

Genetics is a biological science. Science is neutral. But when genetics is used to study the population history of Afro-Americans and Africans it can become another tool to promote White Supremacy.

Admixture research studies support white supremacy, because they are based on Eurocentric historical studies that do not reflect the actual history of Black and African people. Geneticists have separated humanity into four main groupings: Africans, Europeans, Asians and Native Americans.

9

The members of these groupings can be consider a clan.

The members of each grouping have been assigned varying haplogroups that are supposed to be characteristic of that group or population. Each haplogroup has subclans called subclades for the Y-DNA and mtDNA haplogroups.

Genetics research like white people writing history is used to control Afro-Americans and is based on lies. Geneticists assume that you can discuss the genetic history of Afro-Americans as a simple case of admixture resulting from the mixture of Africans, Europeans and Native Americans since the Transatlantic Slave Trade. This is false.

Genetic admixture studies are invalid and unreliable because Africans or Negroes have been in contact with Native Americans and whites or Europeans for 1000s of years. Archaeogenetic evidence of this reality will be discussed in the upcoming chapters of this book.

As a result, of the long history of contact between Blacks, Europeans and Native Americans, genetic admixture studies make it appear that Afro-Americans only carry mtDNA of the L macrohaplogroup, and the Y-Chromosomes carried by Africans are supposed to only include A,B, and E. And that any other haplogroups carried by Africans are the result of a back a migration of Europeans or Asians into sub-Sharan Africa.

The results, of admixture studies are invalid because the origin dates for the so-called "European" Haplogroups are all before Anatomically Modern Humans replaced Neanderthals in Europe and the Levant. Standard population genetics papers hypothesize that many so-called Eurasian genes entered Eurasia via the Levant. The major problem with this theory is that Neanderthal's had firm control of the Levant up to 32kya. So, haplogroups L3(M,N) and Y-haplogroup R probably originated in Africa and spread to the Americas

Genetics research is supposed to tell us population history. But genetics cannot tell us the history of admixture between African and non-African groups, because the admixture f these populations precede Columbus by 2000 years. The status quo history used to support genetics research articles of African people is a lie. Since the Transatlantic Slave Trade scholars have failed to write an accurate history of Black and African people due to White Supremist ideas that claim Africans don't have a history except as slaves on Americas plantations in the Old South.

Amos Wilson in **The Falsification of Afrikan consciousness: Eurocentric history, psychiatry and the politics of white Supremacy** believes that the African spirit and mind can be healed through the advancement of African centered historiographic, social and natural sciences. Wilson wrote "Apparently the rewriting, the distortion and the stealing of our history must serve vital economic, political and social functions for the Europeans or else he would not bother and try so hard to keep our history away from us, and to distort it in our own minds" (p.15).

Dr. Wilson adds that "We must recognize the intimate relationship between culture history and personality. If we do not know our history then we do not know our personality".

To Wilson we should see history as psychohistory, since the aim of writing Black people out of history is to destroy any sense of intellectual or social self-esteem for African people. Wilson noted that" In the final analysis, European history's principal function is to first separate us from ourselves and separate us from the reality of the world; to separate us from the reality of our history and to separate us from its ramifications"(p.24).

Wilson maintains that we must study Afrocentric History, because Europeans use history as a way of maintaining white supremacy; and the study of history by Blacks is a threat to the status quo. Some Black people believe that the writing of history is neutral. Writing history is not neutral.

Michael Parenti, in History as Mystery (1999), believes that history is not neutral. In his opinion history is written by the ruling class to solidify their position. He observed that "much written history is an ideologically safe commodity. It might best be called "mainstream history", "orthodox history", "conventional history" and even "ruling-class history" because it presents the dominant perspective of the affluent people who preside over the major institutions of society" (Perenti, p.xi).

Parenti, supports Wilsons' view on the impact of Eurocentrism on education when he noted that " many history and political science programs offered in middle and higher education rest on a Eurocentric bias" (p.xiv). As a result, Parenti argues that we learned a "disinformational history" which represents the views of the ruling class rather than real history (p.10). As a result, Parenti claims that we have "consensus history textbooks" that teach history from a distorted base.

The comments of Wilson and Parenti, make it clear that history is not written from a neutral perspective, it is written by historians who define what history is or is not. This means that due to doxa, the personality and preconceptions of the historians determine how he writes history. As a result, we find that "establishment" historians usually write history which supports the dominant view of the ruling class, which primarily support institutions of higher learning through well-funded endowments. The allegiance of a particular historian to a class or "association" means that when the historian identifies, selects and interprets facts, and the framework used to appraise the facts will be guided by the truths accepted by the "association" or social class.

This is why Jacques Berlinerblau, in **Heresy in the University: The Black Athena controversy and the responsibilities of American intellectuals** (1999), observed that "How can a social-scientist, a historian, a literary criticism etc., claim that his or her conclusion are in any way true when it is so abundantly clear that these conclusions are inextricably bound with the social and political contexts in which he or she works and lives?"(p.192).

Geneticists are not intentionally promoting white supremacy. The white supremacy comes from the fact that they use "consensus history textbooks", that fail to accurately record the history and archaeology related to Black civilizations.

Population Genetics is also used to support white supremacy. The Structure and Admixture programs are used by geneticists to determine genetic admixture.

The Structure and Admixture programs assume that the transatlantic slave trade, marks the first contact between Africans, Europeans and Native Americans. This assumption of post Columbus, first contact for these populations allows researchers to make inferences when comparing Afro-Americans to non-Afro-Americans. Research studies based on this theory almost always have the same result, partial European ancestry and a small fraction of Native American ancestry.

In reality, these modeling programs cannot tell us anything about the genetic history of Afro-

Americans because West Africans have been in contact with whites and Native Americans for thousands of years as you will learn from the upcoming chapters of this book.

Finally, genetic admixture studies make it appear that Afro-Americans only carry mtDNA of the L macrohaplogroup, and the Y-Chromosomes carried by Africans are suppose only include A,B, and E. And that any other haplogroups carried by Africans are the result of a back a migration of Europeans into sub-Sharan Africa. The result of admixture studies are invalid because the origin dates for the so-called "European" Haplogroups are all before Anatomically Modern Humans replaced Neanderthals in Europe and the Levant.

In summary, population genetics and admixture studies lacks a foundation because Africans have been in contact with Native Americans and Europeans for thousands of years before Columbus found the Americas. As a result, basing phylogeography on the myth, Africans met whites only during the Transatlantic slave trade invalidates the results of their studies and speculation on what haplogroups are carried by particular continental populations.

Although admixture studies cannot validate which

population originally carried this or that haplogroup, these studies can tell us which population carries a particular haplogroup today.

This book will explore the prehistory of African and Afro-American populations through phylogeography. Phylogeography has been defined as the study of the historical mechanisms that might be accountable for the present geographical distributions of individual populations using population genetics. Population genetics is a branch of biology interested in the hereditary makeup of a population(s).

This book will provide a short introduction to African and Afro-American population genetics. It includes a chapter on the status of phylogeographical research in Africa

There are four main groupings of humanity: Africans, Europeans, Asians and Native Americans. The members of these groupings can carry varying haplogroups can be consider a clan. Each haplogroup has subclans called subclades for the Y-DNA and mtDNA haplogroups.

We will discuss the demographic history of Africans and Afro-Americans. Afro-Americans can be defined as the dark pigmented ,yellow to copper colored people in the United States whose ancestors were Black Native Americans, Black Europeans and Sub-Saharan Africans taken to the United States during the TransAtlantic Slave Trade. The diverse origins of the Afro-American (AA) population, makes the AA population heterogeneous indigenous people of the United States. Due to the Black Native American origins of the AA people they can be considered an indigenous American population who very phenotypically.

The African and Afro-American populations contain considerable heterozygosity because of the diverse origins of these populations and gene flows from varied geographical regions in Africa and the Americas, leading to the raise of Afro-American and African populations. For purposes of this we will discuss only mtDNA and Y-chromosome markers, primarily haplogroups. This will allow us to look at the rich genetic heritage of the AA and Sub-Saharan African (SSA) populations.

DNA carries genetic information that is passed down from mothers and fathers to their daughters and sons from generation to generation. We inherit mtDNA from our mothers and Y-chromosome DNA from our fathers.

The mtDNA and Y-chromosome are placed in individual groups or branches we call haplogroups. The haplogroup asignments trace your ancestors back to a single individual. The members of the varied haplogroups represent your maternal line and paternal line.

A haplogroup is similar to a large family that share specific characteristics unique to that family. There are two families of haplogroups mtDNA and Y-chromosome DNA.

Haplogroups are assigned a capital letter and a number to designate a branch of a particular mitochondrial or Y-chromosome DNA tree. Each haplogroup has subclades that represent the common ancestry of individuals.

The names of haplogroups are arbitrary. They usually begin with a capital letter that has a number attached to it, e.g., R1. To distinguish subgroups of a lineage lower case letters and numbers are assigned to haplogroups, e.g., R1b1a.

Each Haplogroup is assigned an age. The age of a Haplogroup is an approximate age. The age assigned to a Haplogroup is not an exact age for that Haplogroup it is just an estimate based on statistical modeling.

Mitochondrial DNA (mtDNA) is related to the maternal or female side. The mtDNA is passed on from the parent to their offspring. The mtDNA provides information on the maternal line of the individual. Thusly maternal haplogroups are haplogroups passed down from the mother.

The Y-chromosome DNA is passed from father to son only. This is because only males possess the y-chromosome. As a result, the Y-chromosome represents the paternal line.

I have written this book because there is a demand for a book that tells the phylogeography of African people from an African centered perspective.

It is sad, population genetics are rarely discussed by African people.

There are a number of professional Black anthropologists out there but they are not writing on these themes. They prefer to write on American slavery including "excavating slave sites"--instead of researching all aspects of Black history.

To understand these phenomena you have to study the tradition of historiography practiced by Afro-Americans. Traditionally, AA academics have been cowards--even when they taught at the Negro Colleges.

A good example was Leo Hansberry. Even though Hansberry taught at Howard University and was the first American professional historian to write on African history—he was not made a member of Howard University's African Studies Department.

As a result, in the 20th Century beginning with W.E,B. DuBois, independent researchers have written ancient Black history, not professional anthropologists and historians. This is why modern studies of ancient African and Black civilizations begins with DuBois' **The Negro**.

Sometimes population geneticists publish all their research. When you look at the data you can find data which support an alternative interpretation of there results. For example, Kivisild published a 1999 article where he found hg M1 in India.

Later the Hindu nationalists came to power and Indian geneticists began to spread the myth the Dravidians were not related to Africans—but the 1999 article proves them wrong along with HLA and y-chromosome data.

The Hindus are descendants of the Aryans. We know they don't enter India until after 900 BC. So geneticists have attempted to claim the Dravidians developed in situ, and that since IE speakers share many Dravidian genes—there was no Aryan invasion of India and therefore Indian civilization was founded by Aryan people.

The Dravidian carry hgs M and R. If Dravidians originated in Africa and only recently came to India only 4.5kya , Africans directly influenced IE speaking people. You take away the Dravidian connection to hgs M and R, hgs M and R become African genes, and prove that they developed in Africa—not Asia.

Trying to deny the African origin of the Dravidians is a waste of time. A well founded hypothesis leads to new hypotheses. Since the classical writers it was known that Indians and Africans were related. As a result, you can't change this truth.

This means that researchers writing population genetics from an African centered perspective are isolated. This may change in the future. The introduction of archaeogenetics, as a method of research in population genetics may have a great influence on the field since it encourages the use of linguistic, archaeological, and craniometric data to support the genomic evidence.

This is an important development because most population genetic studies lack any collateral data to support the genetic data. The dating for many demographic movements is based solely on statistical models—not archaeology. As a result they can not meet the standard.

The first thing I learned from my PhD advisor is that good hypotheses are self-generating. This is why geneticists will never be able to maintain the myth of an in situ origin of Dravidian speakers.

This is why Kivisild was upset that I found the 1999 article where they admit to finding M1 in India.

F. Cruciani is a master tactician of the Eurocentrists and the myth of a back migration introducing haplogroups M and R to Africans. He made M173 in Africa R*-M173. The addition of the * , made R*-M173 a paragroup—which means that it needs further elaboration of its location in the hg R tree. This is why it was easy to give R*-M173 the designation V88 , and give it a new name to separate it from Europeans carrying the same gene.

Cruciani et al (2010) work fails to support an Asia origin for hg R. Eventhough he created R-V88, to include all the African R-P25s you still have other African populations that carry R-M269.

The finding of the V45 mutation supports an African origin for hg R. This results, from the fact that ISOGG 2010 y-DNA V45 is phylogenetically equivalent to M207. The presence of V45 makes it clear that hg R exist in Africa and the diversity of R haplogroups (hg) in Africa is due to the African origination of this gene.

The only way you get the truth out is to publish articles disconfirming the research of the Hindu nationalists and Eurocentric geneticists.

# Chapter One: The Major African and Afro-American Haplogroups

Today scientist use genetics to group people into populations. The mtDNA is used to determine the female line of AMH; while y-chromosomes are used to determine your male ancestors.

The Pan-African haplotypes are 16189, 16192, 16223, 16278, 16294, 16309, qnd 16390. This sequence is found in the L2a1 haplotype which is highly frequent among the Mande speaking group and the Wolof.

Haplotypes with HVSI transitions defining 16129-16223-16249-16278-16311-16362; and 16129-16223-16234-16249-16211-16362 have been found in Thailand and among the Han Chinese (Fucharoen et al., 2001; Yao et al., 2002) and these were originally thought to be members of Haplogroup M1. However, on the basis of currently available FGS sequences, carriers of these markers have been found to be in the D4a branch of Haplogroup D, the most widespread branch of M 1 in East Asia (Fucharoen et al., 2001; Yao et al., 2002). The transitions 16129, 16189, 16249 and 16311 are known to be recurrent in various branches of Haplogroup M, especially M1 and D4.

25

There is mtDNA data uniting Africans and Dravidians. Some researchers attempt to portray the Dravidians as Caucasoid people and try to link these people to western Eurasian populations. Other researchers in India attempt to postulate an Indian origin for Dravidians because they mainly belong to the M haplogroup (HG).

The most ancient haplogroups are carried by the Khoisan and Pygmy people. These haplogroups include L1, A, B and etc. The Khoisan mtDNA was named originally L1a,L1d and L1k, these clades are called LoD and LoK today.

The mtDNA haplogroups L1, L2, L3 and U5 are associated with Niger-Congo speakers. Phylogenetically all the Eurasian mtDNA branches descend from L3.

The Khoisan carry L1c,L1i, L2b, L3d ( Rito, et al ,2013) . The motif L3b, is widespread in western Africans. It is mainly found among populations that speak languages of the Niger-Congo family like the Mandekan.

Like most African haplogroups the control region of hg L1i include 16189,16223 and 16311, just like L3a and L3b. The mutation that connects the Khoisan to the spread of L3(M,N) is AF24. The AF24 mutation is found in LOd and among the Khoisan and Senegalese .The existence of AF24 in Senegal and Southern Africa suggest that L1c, L2b, L3d and L3e is not the result of intermarriage with Bantu immigrants , as suggested by

Rito et al(2013).

Frigi et al (2010), in Ancient Local Evolution of African mtDNA Haplogroups in Tunisian Berber Populations noted that: "Our results also point to a less ancient western African gene flow to Tunisia involving haplogroups L2a and L3b. Thus the sub-Saharan contribution to northern Africa starting from the east would have taken place before the Neolithic. The western African contribution to North Africa should have occurred before the Sahara's formation (15,000 BP)".

This would explain why Pericot and Dominguez (2005) found evidence of hg L3 at ancient Iberian sites. Luis Pericot was sure that the populations associated with the Gravettian (32kya) and Soultrean (23kya) cultures were phylogenetically Sub-Saharan African (Dominguez,2005). Dominguez (2005) found that the lineages recovered from ancient skeletons associated with these cultures belonged to the African lineages L1b,L2 and L3. Almost 50% of the lineages from the Abauntz Chalcolithic deposits and Tres Montes, in Navarre are the Sub-Saharan lineages L1b,L2 and L3.

LOd is the oldest mtDNA haplogroup. This haplogroup is primarily carried by the Khoisan people (Winters,2014). It is also found among Niger-Congo speakers in West Africa where we also find LOa in West Africa in addition to L3b.

Haplogroup LOd is found at the root of human mtDNA. Gonder et al (2006) maintains that LOd is "the most basal branch of the gene tree". The TMRCA for LOd is 106kya. This makes haplotype AF-24 much older than L3a and probably explains why this haplotype is found among the Khwe/Khoisan (Chen et al, 2000).

The TMRCA of LOd dates to 106kya. As a result, anatomically modern humans (amh) had plenty of time to spread this haplogroup to Senegal. In West Africa the presence of amh date to the Upper Palaeolithic (Giresse, 2008). The archaeological evidence makes it clear that amh had ample opportunity to spread LOd and L3(M,N) which has an affinity to AF-24 (Chen, 2000), to West Africa during this early period of demic diffusion of amh in Africa.

The earliest evidence of human activity in West Africa is typified by the Sangoan industry (Phillipson, 2005). The amh associated with the Sangoan culture may have deposited Hg LOd and haplotype AF-24 in Senegal thousands of years before the exit of amh from Africa. This is because it was not until 65kya that the TMRCA of non-African L3(M,N) exited Africa (Kivisild et al, 2006).

Anatomically modern humans arrived in Senegal during

the Sangoan period. Sangoan artifacts spread from East Africa to West Africa between 100-80kya. In Senegal Sangoan material has been found near Cap Manuel (Giresse, 2008), Gambia River in Senegal (Davies, 1967; Wai-Ogussu, 1973); and Cap Vert (Phillipson, 2005).

The presence of the AF-24 is a haplotype of haplogroup LOd makes it clear that this haplotype is not only an ancient human genome. It is also evidence that AF-24 probably did not originate in Asia, since AF-24 was found among the Senegalese and Khoisan.

This reflects an early migration from East Africa to West Africa. The presence of basal nucleotides characteristic of macrohaplogroup L3(M) in West Africa and the reality that M1 does not descend from an Asian M macrohaplogroup because of the absence of AF24 in Asia (Sun et al, 2005) and its presence among the Khoisan and Senegalese suggest that expansion of M1 was probably from Africa to Eurasia. The existence of haplotype AF-24 and basal L3(M) lineage in East and West Africa suggest the probable existence of the Proto-M1 lineage in Africa, not Eurasia before haplogroup L3(M,N) carriers exited Africa.

Atkinson et al (2009) makes it clear that L3 is the youngest African haplogroup and LO is the oldest. As a result, when Dr. Oppenheimer claims that LOd is not the TMRC of AMH, he is false. LO is the oldest haplogroup, since LOd is dated to 106kya and one of the LO clades it. Atkinson et al (2009) observed that "Haplogroups LO and L1 (figure 2b,c, respectively) show slow constant growth over the last 100–200 kyr (TMRCAs: LO, 124–172 kyr ago; L1, 87–139 kyr ago; LO and L1 combined, 156–213 kyr ago; 95% HPDs)". This makes it clear that haplogroup LO is the oldest mtDNA haplogroup in Africa.

This would explain why Pericot and Dominguez (2005) found evidence of hg L3 at ancient Iberian sites. Luis Pericot was sure that the populations associated with the Gravettian (32kya) and Soultrean (23kya) cultures were phylogenetically Sub-Saharan African (Dominguez,2005). Dominguez (2005) found that the lineages recovered from ancient skeletons associated with these cultures belonged to the African lineages L1b,L2 and L3. Almost 50% of the lineages from the Abauntz Chalcolithic deposits and Tres Montes, in Navarre are the Sub-Saharan lineages L1b,L2 and L3.

Dr. Oppenheimer also claims that haplotype AF-24 is " poorly resolved". This is false, Chen et al make it clear that" The samples included complete haplotypes of 62 Senegalese (AF01–AF24, AF26–AF36, AF45–AF59, AF64–AF65, and AF70–AF79)". As a result, how can he make the claim AF-24 is poorly resolved when Chen et al (2002) make it clear that he used "complete haplotypes of 62 Senegalese" samples that include AF-24.

Chen et al makes it clear that AF-24 could be of either Asian or African origin"Similarly, L3a was found to have a close affinity to haplotype AF24, a mtDNA that has the Ddel np-10394 and Alul np-10397 site gains characteristic of Asian macrohaplogroup M. Therefore, it is possible that subhaplogroup L3a was the progenitor of Asian mtDNAs belonging to M. Although the age of subhaplogroup L3a is somewhat less than our estimate for the age of Asian haplogroup M (Torroni et al. 1994b; Chen et al. 1995), the differences could be due to the limited number of L3a mtDNAs in our African sample. Alternatively, AF24 may have been introduced from Asia into Africa more recently." The fact that Atkinson et al (2009) makes it clear that AF-24 is a haplotype of LO, make it unlikely that AF-24 originated in Asia, since it was already in existence prior to the OoA event.

Finally, Oppenheimer claims that you can not infer population movements relating to the expansion of the ancient tool kits. This is a false statement since the demic expansion of LO(d) and L3 from East Africa to West Africa is cross referenced with specific founding lineage which is assumed to have originated in the East. This assumption is just as valid as Oppenheimer's view relating to the Tonga event's impact on the OoA.

It is obvious that Dr. Oppenheimer has little knowledge of the expansion of haplogroups in Africa. I am surprised the he didn't know that the GenBank Accession number for Haplotype AF-24 is DQ112852, this suggest that he is not keeping up with the literature. Moreover, the earliest examples of L3(N) come from Iberia, not East Asia. Since this area was first occupied by Neanderthals until the expansion of the Aurignacian culture which had to have crossed the Straits of Gibraltar from Africa (Winters, 2012). No where in Dr. Oppenhiemer's response dose he present textual evidence supporting his conclusions. He only provides his opinions—not evidence.

## Haplogroup U

Sanchez-Quinto et al analyzed shared ancient genome of European farmers 9kya that indicated these hunter-gatherers carried U5 and were in communication from Iberia to Central Europe. These researchers are correct to emphasize the unity between these geographically diverse groups .

Sanchez-Quinto et al suggest that the U5 lineage probably originated in Central Europe and evidence a back-to-Africa migration. This hypothesis lacks continuity because the populations with the highest frequencies of the U5 cline are found either in Northwest Africa (NWA) or Sub-Saharan West Africa (SSWA).

A phylodemographic analysis of contemporary Europeans indicate that only 4%-7% Europeans carry the U5 lineage. This discontinuity of the European genetic landscape manifest questions on the ancestry of the ancient European hunter-gatherers. The genetic data would make the source of U5 in the western parts of Africa—not Eurasia.

Malyarchuck et al believes that the U5 lineage arrived in Europe with the Aurignacian culture bearers. This seems highly unlikely because the Aurignacians carried mtDNA L3(N) when they crossed the Straits of Gibraltar 40kya . As illustrated by Haak et al , ancient mtDNA indicates that hunter-gatherers belonging to the N1a haplogroup was the predominate genome among the Aurignacians up to 9 kya . Consequently, the genetic evidence suggest a replacement of the Aurignacians by a populations carrying the U5 clades.

The closest European population to the ancient European hunter-gatherers are the Saami of Scandinavia. The Saami carry the highest frequencies of the U5 lineage . The predominate Saami U5 clade is U5b1b1. Whereas only 4% of contemporary Europeans carry U5, 50% of the Saami belong to U5b.

The highest concentration of U5 is found among Berbers in NWA . It is also carried by Mande and Fulani Niger-Congo speakers in West Africa (5-8).

The U5 haplogroup carried by the Mande, like other SSWA is characterized by 16189,16192,16270 and

16320. There is a high frequency of U5 among the Fulani who mainly belongs to U5b1b and U5a . Interestingly, U5b1b links the Saami of Scandinavia and the Yakuts to the Fulani and NWA Berbers .The frequency of U5 in SSWA and NWA suggest demic diffusion of these populations across the Straits of Gibraltar into Iberia and thence Central Europe.

Achilli et al has argued that there was a back migration of the U5 cline across the Straits of Gibraltar. This hypothesis lacks congruence given the early date of U5 in Iberia, in comparison to Central Europe; and the presence of diverse sub-clades of U5 distributed across SSWA among diverse populations which fail to record a history of mating with Berber populations. Moreover it is clear that groups such as the Fulani are of African origin and show no history of admixture with Europeans.

Rosa et al claims an autochthonous origin for haplogroup U6 in NWA given the diversity of U6 clades in NWA around 38kya. Using this criterion, the diversity of U5 clades in SSWA and the phylogeography of U5 in this region, support an in situ origin for the U5 clade in the same region as haplogroup U6.

The genetic data from contemporary European populations fails to support a migration of populations carrying haplogroup U5 into Europe via the Levant. The low frequency of U5 in Europe, except among the Saami, probably indicates a single episode of ancient gene flow from NWA and SSWA into Iberia 9kya. The present phylogeograpical distribution of the U5 cline reflects demographic porcesses involving population replacement, drift and a history of genetic bottlenecks resulting from demic diffussion of neolithic populations from the Levant and Central Asia into Europe.

The temporal and spatial distribution of U5 and U6 clades outside Europe, point to a NWA and SSWA origin for these lineages. The high frequency of U5 in NWA and SSWA suggest the spread of the U5 cline into Iberia across the Straits of Gibraltar 7kya.

### Haplogroup L3(M,N)

Dr. Oppenheimer (2012b) implies that L3(M,N) originated in Asia. This is false. We know that L3 originated long before the OoA event. He does not present any evidence falsifying my conclusion. His entire argument is that M1 is 'rare' in Asia.

L3(M,N) probably spread across Africa before the Out of Africa event that led to the dispersal of anatomically modern humans to Eurasia. Sores et al (2012) admit that the several branches of L3 probably expanded soon after the emergence of L3, they fail to comprehend the full extent of this finding. If we are using the Toba super-eruptiobn of 73.5ka as a baseline blocking any OoA event of amh (Oppenheimer, 2012; Sores et al, 2012), L(M,N) must have spread across Africa before the OoA event.

Sores et al (2012) believes that mtDNA haplogroups M and N originated in Eurasia. Sores et al (2012) founded this conclusion on Olivieri et al (2006) who maintained that around 50kya, M1 entered Africa as a result of a bacxk migration from Eurasia; and a sample limited to Sudanese, Ethiopians and Somalis.

Sores et al (2012) argues that L3(M,N and R) originated in Eurasia, based on Olivieri et al (2006). Olivieri et al (2006) suggest that haplogroup M1 probably did not originate in Africa before the out of Africa exit. They opine that M1 probably represents a back-migration into Ethiopia. Olivieri et al base this conclusion on :

1) the absence of any distinguishing M1 root mutations in Asian M haplogroups ;

2) the presence of M1 only in East Africa and North Africa; and

3) the lack of any Asian specific clades within M1

Olivieri et al (2006) argue that M1 was probably spread from the Levant back into Africa by the Aurignacian culture. The craniofacial and molecular evidence does not support this conclusion (Winters, 2007,2008b).

Gonzalez et al (2007) and Olivieri et al (2006) have assumed that haplogroup L3(M,N), was probably carried to western Eurasia via the Levant. But the archaeological and craniometric eveidence indicates that the Levant was still occupied by Neanderthal man until 32kya. The archaeological evidence indicates that Around 40kya Europe was still occupied mainly by Neanderthals.

The archaeological record informs us that Cro-Magnon people belonging to the Aurignacian culture carried hg N (Caramelli et al, 2003). They replaced the Neanderthal population of the Levant, at Ksar Akil around 32, 000 years ago (Gilead, 2005; Steven, 2001). This was 10k after Cro-Magnon people had settled Iberia.

Anatomically modern humans replaced in Europe around 32,000 by the CroMagnon people at Les Eyzies in France. It is also evident that archaic humans were replaced in much of the Levant by the Levantine Aurignacian culture bearers by a local variant of this technology at Ksar Akil XIII-VII 32kya, not 60-50kya. The archaeological evidence makes a back migration from the Levant.

Oppenheimer (2012) makes it clear that amh colonization of the Levant failed; and by60kya Neanderthal populations dominated the Levant.

Clearly, the dates for L3(M,N) in western Eurasian are incongruent to TMRCA of the populations carrying the L3(M,N) lineages into eastern Eurasia which probably date to 60-65kya. This incongruence in relation to the dates for this haplogroup in eastern Eurasia, and its complete absence in much of western Eurasia today suggest that the population carrying this gene into Eurasia may not have entered Eurasian during the recognized Africa exit event.

There is considerable evidence that M1 is found in Asia. Researchers have found the M1 haplogroup in the Caucasus ( Bermisheva et al, 2004; Tambets et al, 2000), Central Asia, and East Asia (Comas et al, 1998; Fucharoen et al, 2001). In addition, the Russian haplotype 16183c-16189 ,16249, 16311 match the M1 HVSI sequence (Malyarchuk et al, 2004).

**Even though Olivieri et al (2007) claim that East African** M1 root mutations are absent in Eurasian M sister clades is not supported by the evidence. For example researchers have found that the Tanzanian M1 haplogroup cluster with people from Oceania (Gonder et al, 2006). And, as mentioned earlier the M1 mutations 16129,16189,16249 and 16311 are found in many southeast and East Asian haplogroups (Fucharoen et al, 2001; Yao et al, 2002).

It is also not true that HG M1 is absent in India. Kivisild et al noted that 26 of the subjects in their study belonged to the M1 haplogroup. These researchers reported sub-cluster M1 was found mainly in Kerala and Karnataka high caste individuals.

It is clear that the molecular evidence does not support Olivieri et al (2006) hypothesis that M1 is probably the result of a back migration. This evidence on the other hand confirms the hypothesis of Quintana-Murci et al (1999) that M1 was probably already present in East Africa when the out of Africa exit/ event took place.

The molecular evidence makes it clear that haplogroup M1 is not confined solely to Ethiopia as maintained by Olivieri et al (2006). This haplogroup along with HGs N and M*, are also found in Tanzania, Uganda, Egypt and the Senegambian region (Gonzalez et al, 2006; Gonder et al, 2006; Winters, 2007). In Tanzania the predominate M1 clades are M1 , M1a1 and M1a5. In Senegal the predominate M1 lineage is M1c1.

In addition to M1 in Africa, we also find haplogroups M*, M23, M3 positions 482 and 16126; M30 positions 195A and 15431; and M33 position 2361. It is interesting to note that the presence of these genes, which are normally found in India are also found in

Africa, is interesting given the presence of M1 in India and the existence of these genes among populations stretching from Africa into Yemen on into India along a path associated with the spread of the Tihama culture (Winters, 2008) .

In addition to haplogroups M1, M* and N in Sub-Saharan Africa we also find among the Senegambians hapotype AF24 (DQ112852) , which is delineated by a Ddel site at 10394 and Alul site of np 10397. The AF-24 haplotype is a branch of the African subhaplogroup L3 (Chen, 2000). This is the same delineation of haplogroup M*. It is clear from the molecular evidence that the M1, M and N haplogroups are found not only in Northeast Africa, but across Africa from East to West (Winters, 2007).

Neanderthal dominated the Levant when the imagined back migration of M1 occurred 50kya ,we must reject the contention of Gonzalez et al. (2007) and Olivieri et al. (2007) and Sores et al (2012) that M1 originated in Asia because 1) the possible Senegalese origin of the M1c subclade; 2) the absence of the AF-24 haplotype of haplogroup LOd in Asia; and 3) the African origin of the Dravidian speakers of India (2007,2008)who carry the most diverse M haplogroups.

Moreover, the existence of the L3a(M) motif in the Senegambia characterized by the Ddel site np 10394 and Alul site np 10397 in haplotype AF24 (DQ112852) make a 'back migration of M1 to Africa highly unlikely, because of the ancientness of this haplotype. The first amh to reach Senegal belonged to the Sangoan culture which spread from East Africa to West Africa probably between 100-80kya.

Thangaraj et al recognize a Paleolithic origin for the M haplogroups in India. The majority of Dravidian speaking people belong to the M haplogroup. Most geneticists agree that the M macrohaplogroups are derived from L3. Kivisild et al made it clear that all Indian mtDNA lineages "coalesce finally to the African L3a".

Metspalu argues that the earliest offshoots for L3, were HGs M and N developed in Arabia. Metspalu believes the MRCA for the M HG entered Asia 60-65 kya .

Metspalu maintains that "all the basal trunks of M, N and R have diversified in situ" (p.24). He makes it clear that in his opinion the M HGs are different from the subhaplogroup M of East Asia . The most frequent HG in India is M2.

Sixty percent of of the Indian mtDNA lineages are M HGs . Kivisild et al maintains that there are five M HGs in India: M1, M2,M3, M4, and M5. Thanaraj et al [42] has revised the classification of HGs M3, M18 and M31 and defined the novel HG M41. Sun et al [45] identified another 5 M HGs (M34-M40) in addition to the Indian mtDNA macrohaplogroup N.

The diversity of M HGs in India has led many researchers to suggest that the M clades have an in-situ origin .These researchers speculate that although L3 originated in Africa, the M1 HG in Ethiopia and Egypt ,may be the result of a back migration to Africa from India .

These researchers base this theory for a back migration to Africa from India, on 1) HG M1 is not found in India; and 2) the MHG's are only found in East Africa . Both of these theories have little support when we look at the mtDNA data for Africa and India.

Barnabas et al noted that N,M and F lineages found in India could have originated in Africa (pp.13-14). He speculated these people migrated to India from Africa during the Upper Paleolithic.

Most researchers make it appear that the M1 haplogroup is only found in Ethiopia. These researchers maintain that the M1 HG is restricted to the Afro-Asiatic linguistic phylum. This is false M HGs are found in other parts of Africa where people speak non-Afro-Asiatic languages.

The M lineages are not found only in East Africa. Rosa et al [46] found a low frequency of the M1 HG among West Africans who speak the Niger Congo languages, such as the Balanta-Djola. Gonzalez et al found N, M and M1 HGs among Niger-Congo speakers living in Cameroon, Senegambia and Guinea Bissau.

It is also not true that HG M1 is absent in India. Kivisild et al [41] found five M HGs in India: M1, M2, M3, M4 and M5. It is interesting to note that the M4 HG has the same 16311 coding region as the African M1 HG.

Kivisild et al provides the first detailed discussion of the M subclusters in India and suggested an autochthonous development of these lineages in India. The researchers suggest that there were multiple M lineages when this haplogroup migrated to Asia. These researchers claimed that the expansion date for the five M subclusters expanded into India between 17,000-32,000 bp.

Kivisild et al (1999) noted that 26 of the subjects in his study belonged to the M1 haplogroup. It is clear from this study that sub-cluster M1 was found mainly in the Indian states of Kerala and Karnataka. An interesting finding in the study was that most of the Indians with the M1 HG were members of upper caste. Africans and Dravidians share haplogroups M1, M3, M30 and M33.

**Phylogeography of y-Chromosome**

The phylogeography of y-Chromosome haplotypes shared among the Niger-Congo speakers include A,B, Elb1a, E1b1b, E2, E3a and R1 . The predominate y-Chromosome among the Niger-Congo is M2, M35, and M33.

Haplogroup E has three branches carried by Niger-Congo populations E1, E2 and E3. The E1 and E2 clines are found exclusively in Africa. Haplogroup E3 is also found in Eurasia. Haplogroup E3 subclades are E3b, E-M78, E-M81 and E-M34.

The majority of Niger-Congo speakers belong to E1b1a, E1b1b, E2 and R1. Around 90% belong to y-Chromosome group E (215,M35*).

Y-Chromosome haplogroup A is represented among Niger-Congo speakers. In West Africa, under 5% of the NC speakers belong to group A. Most Niger-Congo speakers who belong to group A are found in East Africa and belong to A3b2-M13: Kenya (13.8) and Tanzanian (7.0%).

Niger Congo speaking populations also carry R1-M269. Gonzalez et al (2012) in a study of R1 in Equatorial Guinea found that 10 out of 19 subjects in the study carried R1b1-P25 or M269. This is highly significant because it indicates that 53% of the R1 carriers in Equatorial Guinea, were M269. Forty-seven percent of the subjects in the study carried V88. This finding is further proof of the widespread nature of this so-called Eurasian gene in Sub-Saharan Africa among populations that have not mated with Europeans.

# Figure 1 Y-Chromosomes A,B, E1a

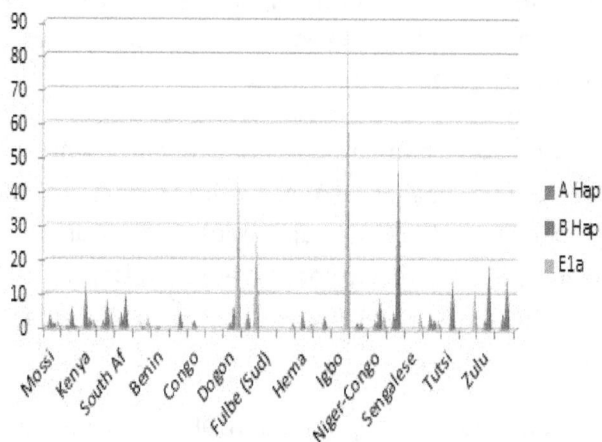

The Bantu expansion is usually associated with the spread of y-Chromosome E3a-M2. In Kenya the frequentcy for E3a-M2 is 52%; and 42% in Tanzania. In Burkina Faso high frequentcies of E-M2* and E-M191* are also represented. It is interesting to note that among the Mande speaking Bisa and Mandekan there are high frequentcies of E-M2*. This is in sharp contrast to the Marka and South Samo who have high frequencies of E-M33.

# Figure 2: Y-Chromosomes E1b1a, E1b1b, E2 and R1

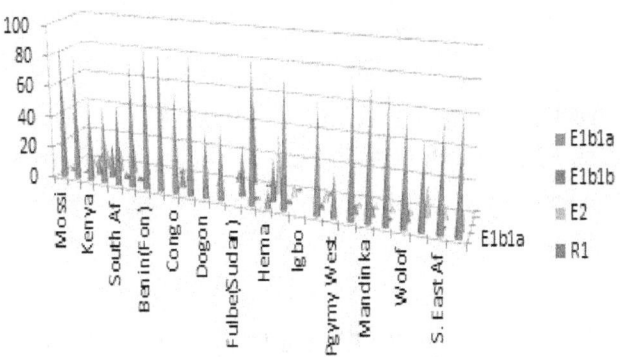

The pristine form of R1-M173 is found in Africa. Y-Chromosome R is characterized by M207/ V45. The V45 mutation is found among NC speakers. The R1b mutations include V7, V8, V45, V69 and V88.

Henn et al presents conclusive evidence that African hunter-gatherer (HG) populations share a number of ancestral lineages including B264*; although they are geographically distinct populations situated among

47

agropastorial groups . An interesting finding of Henn et al was the discovery of the Eurasian clade R1b1b1a1a among the Khomani San of South Africa .

Henn et al was surprised by this revelation of R-M269 among this Khoisan population. As a result, he interviewed the carries of R1b1b2a1a, and learned that no members of their families had relations with Europeans. The presence of R lineages among HG populations is not new. Wood et al reported Khoisan carriers of R-M269 . Bernielle-Lee et al, in their study of the Baka and Bakola pygmies foud the the R1b1* haplogroup . These researchers made it clear that the Baka samples clustered closely to Khoisan samples .

The most common R haplogroup in Africa is V88. Given the interaction between HG groups and agropastoral groups they live in close proximity too, we would assume that African HG would carry the V88 lineage. Yet, as pointed out above the HG populations carry R-M269 instead of V88 . The implications of R-M269 among HG populations, and Henn et al's of shared African HG genome suggest that R-M269 may represent a HG genome.

The low frequency of this Eurasian clade among HG populations may not support this conclusion, tdistribution of R-M269 among HG populations needs further research into the origins of the R y-chromosome among African populations.

The frequentcy of R1-M173 varies among Niger-Congo speakers. The frequency of R-M173 range between 3-54%. The most frequent subtype in Africa is V88 (R1b1a). Haplogroup R1b1a ranges between 2-20% among the Bantu speakers. The highest frequency of R1 is found among Fulbe or Fulani speakers.

Table 1: African Carriers of R-M269

| African Population or Geographical area in Africa | Frequency of Haplogroup RM269 | Reference* |
|---|---|---|
| Africa | 5.2% | Berniell-Lee et al., (2009) |
| Bantu | 2-20% | Berniell-Lee et al., (2009) |
| Pygmy | 5% | Berniell-Lee et al, (2009) |
| Guinea-Bissau | 12% | Carvalho et al., (2011) |
| Equatorial Guinea | 53% | Gonzalez et al., (2012) |
| Khoisan | 2.2% | Wood et al., (2005) |
| Khoisan | 6.0% | Hirbo (2011) |

The phylogenetically deep haplogroup R1b is mainly found in West Africa and the Sahel, where the frequency ranges between 5-85% among various Niger Congo speakers (Cruciani et al, 2010). The paternal record of M173 on the African continent illustrates a greater distribution of this y-chromosome among varied African populations than, in Asia.

The greatest diversity of R1b in Africa is highly suggestive of an Africa origin for this male lineage. Archaeological (Lal, 1963), genetic (Winters, 2008;2010a), placenames (Balakrishnan, 2005) and linguistic data group (Aravanan,1979,1980; Upadhyaya, 1976,1979; Winters 1985a,1985b, 1989) linking Africans and Dravidian support the recent demic diffusion of SubSaharan Africans and gene flow from Africa to Eurasia. An early colonization of Eurasia 4kya by Sub-Saharan Africans and Dravidian carriers of R1-M173 is the best scenario to explain the high frequency and widespread geographical distribution of this y-chromosome on the African continent (Winters, 2010c). Given the greatest diversity of R1- M173, this is the most parsimonious model explaining the frequency of R-M173 in Africa.

The Kushites who belonged to the -roup people of uintroduced R1 to Eurasia (Winters, 2010c).These Kusites founded the Sumerian and Elamite civilizations.

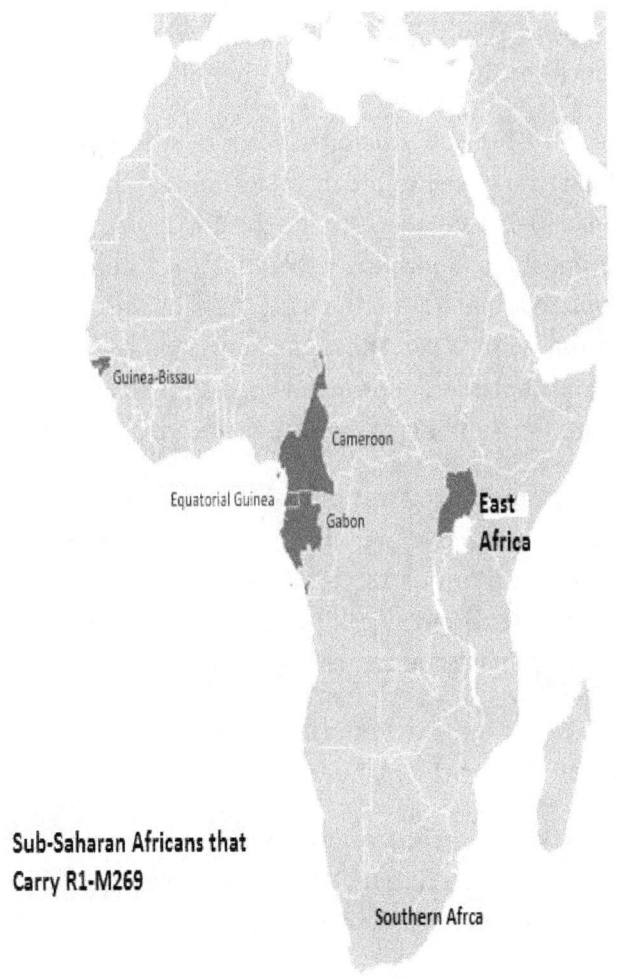

Y-Chromosome R1-M173 was probably spread in

Western Europe first by African Roman soldiers, and later by African Muslims when they conquered Western Europe as Moors. This would explain why 60-70% French and Spanish males carry this y-haplogroup.

Around 0.1 of Sub Saharan Africans carry R1b1b2. Wood et al (2009) found that Khoisan (2.2%) and Niger-Congo (0.4%) speakers carried the R-M269 y-chromosome. Gonzalez, et al (2012), in a study of R1, found that that 53% of the R1 carriers in Equatorial Guinea, were M269. The Niger-Congo speakers formed a significant population in the nomes of Upper Egypt, where the founders of the 18th dynasty originated.

Haplogroup R1b1b2 was probably taken to Europe by African Roman soldiers. Africans were first recorded in the Western Europe 1800 years ago, as Roman soldiers defending Hadrian's Wall. There was a skeleton of an African Roman soilder recently found in Britain.

Other Africans were found in Britain including the Rich African women called the "Bangled lady".

These skeletons show how heavily integrated Africans were in western Europe. This would explain the widespread nature of y-chromosome R1-M173 in Europe.

In addition to R1-M173 in western Europe, the African y-chromosome haplogroup A1 was also recently found in Britain.

We looked at previously published y-Chromosome haplotype gene variants in Indian populations. The H1 haplotype is found among many Dravidians. Sengupta et al [48] noted that the subclades H1 and H2 was found among 26% of the Dravidian speakers in their study, especially in Tamil Nadu . Ramana et al claims that the discovery of H1 and H2 haplotypes among the Siddis is a "signature" of their African ancestry . The frequency of the H1 subclade among Dravidian speakers is also an indicator of an African-Dravidian connection .

In addition to haplotypes H1, in South India we also find the Sickle Cell gene  and African 9-bp deletion . Watkins et al found the 9bp motif among four Indian tribal populations: Irula, Yanadi, Siddi and Maria Gond .

The phylogenetic structure of the Dawoodi Bohra Muslims of Tamil Nadu, India includes African mtDNA and Y-chromosome genes. The Dawodi Bohra carry the mtDNA M1 and Loa2a. The African Y-chromosomes found among the Dawoodi Bohra was 20% haplotype H and 2% E1b1b1a.

Some researchers love to lie and make Black populations into "white" populations. In a recent article researchers claim that the Minoans were white because the majority of Minoans were classified in haplogroups H (43.2%), T (18.9%), K (16.2%) and I (8.1%). Haplogroups U5A, W, J2, U, X and J were each identified

53

in a single individual. As a result, if the majority of Minoans were classified in haplogroups mtDNA H (43.2%) they represent a Black population not white population, since this mtDNA is carried by the Tuareg who are Black.

The Y-chromosomes of Cretans also indicate the Cretans were Blacks. Laisel Martinez et al(2007)[1] provides a detailed discussion of the y-chromosomes in Crete the presence of y-chromosomes R1b, T, K and H in Crete indicate that the Cretans were Black.

Martinez et al (2007), observed that In the case of the R1 haplogroup, while frequencies of 19.2% and 21.7% are found in the Heraklion Prefecture and Lasithi Prefecture populations, respectively, more than half (56.1%) of the Lasithi Plateau individuals are R1-M306-derived.

In the case of Cretan E3b3-M123 (M34) chromosomes, they most likely signal East African or Middle-Eastern gene flow rather than European, due to the scarcity of this lineage in the latter area.19, 26 Similarly, the

---

[1] Laisel Martinez et al , Paleolithic Y-haplogroup heritage predominates in a Cretan highland plateau, Eur J Hum Genet. 2007
http://www.nature.com/ejhg/journal/v15/n4/full/5201769a.html

presence of E3b-M35* individuals in the Heraklion Prefecture population could probably be attributed to an East-African or North-African contribution.

This is interesting because researchers claim that haplotype H indicates that the Siddis, an African population in India are African because they carry haplotype H. Ramana et al (2001) claims that the discovery of H1 and H2 haplotypes among the Siddis is a "signature" of their African ancestry.

The finding that other Minoans carried haplotype T and K also indicates that the Minoans were Blacks, not whites. There are a number of shared African and Indian Y-chromosome haplotypes. These haplotypes include Y-hg T-M70 and H1. Haplogroup T-M70 is found among several Dravidian speaking tribal groups in South India, including the Yerukul (or Kurru) , Gonds and Kols. Y-haplogroup T-M70 is found in the eastern and southern regions of India (Trivedi et al, 2008). It has a relatively high frequency in Uttar Pradesh and Madhya Pradesh (Sharma et al, 2009). Sharma et al (2009) in a study of 674 Dalits found that 89.39 % belonged to Y-hg K*, in relation to Dravidian speakers it was revealed that Y-hg T-M70 was 11.1%. Trevedi et al (2008) report that Y-hg T-M70 is predominately found among Upper Caste Dravidians at a frequency of 31.9. The highest frequency of T-M70 in the World is found among the Fulani (18%) of West Africa. Martinez et al (2007) also found T-M70 and hg K in Crete see the figure above.

Ramana et al (2001) claims that the discovery of H1 and H2 haplotypes among the Siddis is a "signature" of their African ancestry. As a result, the Y-hg H1 subclade frequency among Dravidian speakers can also be considered as an indicator of an African-Cretan-Dravidian connection.

The H1 haplotype is found among many Dravidians. Sengupta et al (2006) noted that the subclades H1 and H2 were found among 26% of the Dravidian speakers in their study, especially in Tamil Nadu. Trivedi et al (2008) found the Y-hg H1 frequency of 22.2 among Dravidian speakers in their study. Sharma et al (2008) reports a frequency rate of 25.2%.

Finally, because the majority of Minoans were classified in mtdna haplogroups H (43.2%), the ancient Minoans were Black, not white, since the Tuareg are Blacks. The presence of y-chromosomes R1-306, R1b, T, K and H in Crete indicate that the Cretans were Black.

In conclusion, the R haplogroup probably originated in Africa. In my paper POSSIBLE AFRICAN ORIGIN OF Y-CHROMOSOME R1-M173 , I argue that the P clade originated in Africa because 1) the age of R-V88 and 2) the widespread nature of R1 in Africa. Researchers have found that the TMRCA of V88 was 9200-5600 kya (Cruciani et al, 2010). Eurasians carry the M269 (R1b1b2) mutation. The subclades of R1b1b2 include Rh1b1b2g (U106) (TMRCA 8.3kya) and R1b1b2h (U152) (TMRCA 7.4kya). The most recent common ancestor for

R1b1b2 is probably 8kya (Balaresque et al, 2010).

In Africa we find R-M269 and V88. Clearly, R-V88 is older than R-M269 , and there is no archaeological evidence of a back migration or haplogroup R into Africa, but there is evidence of the migration of the Kushites and Proto-Saharans into Eurasia from Middle Africa. The diversity of R1 haplogroups in Africa supports the proposition that the R macrohaplogroups originated in Africa, not Eurasia.

**References:**

Achilli A, Rengo C, Battaglia V, Pala M, Olivieri A, et al. 2005. Saami and Berbers – an unexpected mitochondrial DNA link. Am J Hum Genet 76: 883–886.

Atkinson Q D, Gray R D, Drummond A J. 2009. Bayesian coalescent inference of major human mitochondrial DNA haplogroup expansions in Africa. http://rspb.royalsocietypublishing.org/content/276/165 5/367.full

Berniell-Lee, G., Calafell, F., Bosch ,E. ,Heyer, E, Sica, L., Mouguiama-Daouda, | P., van der Veen, L., Hombert, J-M., Quintana-Murci , L.and, Comas, D. (2009) Genetic and Demographic Implications of the Bantu Expansion: Insights from Human Paternal Lineages, Mol. Bio. and Evol. 26(7),1581-1589; doi:10.1093/molbev/msp069.

Carvalho M, Brito P, Bento AM, Gomes V, Antunes H, Costa HA, Lopes V, Serra A, Balsa F, Andrade L, Anjos MJ, Corte-Real F, Gusmão L. (2011).Paternal and maternal lineages in Guinea-Bissau population. Forensic Sci Int Genet. 5(2),114-6.

Cerný V., Hajek M., Bromova M., Cmejla R., Diallo I. & Brdicka R. 2006. MtDNA of Fulani nomads and their genetic relationships to neighboring sedentary populations. Hum. Biol., 78: 9-27.

Chen Y-S., Olckers A., Schurr T.G., Kogelnik A.M., Huoponen K., Wallace D.C. 2000 mtDNA variation in the South African Kung and Khwe - and their genetic relationships to other African populations. Am. J. Hum. Genet., 66, 1362-1383

Coia V., Destro-Bisol G., Verginelli F., Battaggia C., Boschi I., Cruciani F., Spedini G., Comas D. & Calafell F. 2005. Brief communication: mtDNA variation in North Cameroon: lack of Asian lineages and implications for back migration from Asia to sub-Saharan Africa. Am. J. Phys. Anthropol., 128: 678-681.

Comas D, Calafell F, Mateu E, Pérez-Lezaun A, Bosch E, Martínez-Arias R, Clarimon J, Facchini F, Fiori G, Luiselli D, Pettener D, Bertranpetit J (1998). Trading genes along the silk road: mtDNA sequences and the origin of Central Asian populations. Am J Hum Genet, 63:1824-1838.

Cruciani,F., Trombetta,B., Sellitto, D., Massaia,A. destroy-Bisol,G., Watson, E., Colomb, E.B. (2010) Eur J. Hum Genet.,(6 January 2010) doi:10.1038/ejhg.2009.231: 1-8.

Cruciani, F., Santolamazza,P., Shen, P., Macaulay, V., Moral P., Olckers,A. (2002) A Back Migration from Asia to Sub-Saharan Africa is supported by High-Resolution Analysis of Human Y-chromosome Haplotypes. Am J. Hum Genet., 70,1197-1214.

Davies,O. (1967). West Africa before the Europeans. London.

Ely B., Wilson J.L., Jackson F. & Jackson B.A. 2006. African-American mitochondrial DNAs often match mtDNAs found in multiple African ethnic groups. BMC. Biol., 4: 34.

Federico Sánchez-Quinto, Hannes Schroeder, Oscar Ramirez, María C. Ávila-Arcos, Marc Pybus, Iñigo Olalde, Amhed M.V. Velazquez, María Encina Prada Marcos, Julio Manuel Vidal Encinas, Jaume Bertranpetit, Ludovic Orlando, M. Thomas P. Gilbert, Carles Lalueza-Fox. 2012. Genomic Affinities of Two 7,000-Year-Old Iberian Hunter-Gatherers. Current Biology ; DOI: 10.1016/j.cub.2012.06.005

Fucharoen, G., S. Fucharoen and S. Horai, 2001. Mitochondrial DNA polymorphism in Thailand. J.Hum. Genet., 46: 115-125.

Giresse,P. (2008). Tropical and sub-Tropical West Africa—marine and Continental changes during the late Quaternary. Volume 10. Elsevier Science.

Gonder MK, Mortensen HM, Reed FA, de Sousa A, Tishkoff SA.(2006).: Whole mtDNA Genome Sequence Analysis of Ancient African Lineages. Mol Biol Evol., 24(3):757-768.

González, A. M., V. M. Cabrera, J. M. Larruga, A. Tounkara, G. Noumsi, B. N.Thomas and J. M. Mould(2006). Mitochondrial DNA Variation in Mauritania and Mali and their Genetic Relationship to Other Western Africa Populations. Ann of Hum Genet, 70,5. http://www.blackwell-ynergy.com/doi/abs/10.1111/j.1469-1809.2006.00259.x?cookieSet=1&journalCode=ahg

Gonzalez, A. Jose M Larruga , Khaled K Abu-Amero , Yufei Shi , Jose Pestano and Vicente M Cabrera. (2007).Mitochondrial lineage M1 traces an early human backflow to Africa, BMC Genomics , 8:223 doi:10.1186/1471-2164-8-223. Retrieved on 9/15/2010 http://www.biomedcentral.com/1471-2164/8/223

Gonzalez, M. et al. (2012).The genetic landscape of Equatorial Guinea and the origin and migration routes of the Y chromosome haplogroup R-V88. European Journal of Human Genetics advance online publication 15 August 2012; doi: 10.1038/ejhg.2012.167. Retrieved June 7,2016 at:

http://www.ncbi.nlm.nih.gov/pmc/articles/PMC357320 0/

Henn BM, Gignoux CR, Jobin M, Granka JM, Macpherson JM, Kidd JM, Rodríguez-Botigué L, Ramachandran S, Hon L, Brisbin A, Lin AA, Underhill PA, Comas D, Kidd KK, Norman PJ, Parham P, Bustamante CD, Mountain JL, Feldman MW. Hunter-gatherer genomic diversity suggests a southern African origin for modern humans. Proc Natl Acad Sci U S A. 2011 Mar 29;108(13):5154-62. Epub 2011 Mar 7. http://www.pnas.org/content/108/13/5154.full

Holiday, T. (2000). Evolution at the Crossroads:Modern Human Emergence in Western Asia. American Anthropologist, 102(1) : 54-68.

Malyarchuck B, Derenko M, Grzybowski T, Perlova M, Rogalla U et al.2010. The peopling of Europe from the Mitochodrial Haplogroup U5 Perspective. PlosOne 5(4): http://www.plosone.org/article/info:doi/10.1371/journ al.pone.0010285

Oppenheimer S. 2012 Out-of-Africa, the peopling of continents and islands: tracing uniparental gene trees across the map. Phil. Trans. R. Soc. Lond. B, 367, 770-784. (doi: 10.1098/rstb.2011.0306

Oppenheimer, S. 2012b .Response to Winters (2012) 'Haplogroup L3 (M,N) probably spread across Africa before the Out of Africa event'. http://rstb.royalsocietypublishing.org/content/367/159

0/770.full/reply#royptb_el_319

Phillipson, D.W.(2005). African Archaeology. 3rd Edition. Cambridge University Press.

Quintana-Murci L, Semino O, Bandelt H-J, Passarino G, McElreavey K, Santachiara-Benerecetti AS. (1999) Genetic evidence of an early exit of Homo sapiens sapiens from Africa through eastern Africa.Nat Genet 1999, 23(4):437-441.

Rando J.C., Pinto F., Gonzalez A.M., Hernandez M., Larruga J.M., Cabrera V.M. & Bandelt H.J. 1998. Mitochondrial DNA analysis of northwest African populations reveals genetic exchanges with European, near-eastern, and sub-Saharan populations. Ann. Hum. Genet., 62: 531-550.

Rosa A, Brehem A. 2011. African human mtDNA phylogeography at-a-glance. J. Anthropol. Sci, 89:25-58.

Soares P., Ermini L., Thomson, N., Mormina M., Rito T., Rohl A., Salas A., Oppenheimer S., Macaulay V., Richards M.B. 2009 Correcting for purifying selection: an improved human mitochondrial molecular clock. Am. J. Hum. Genet., 84, 740-759. (doi:10.1016/j.ajhg.2009.05.001)

Soare P, Farida Alshamali, Joana B. Pereira, Verónica Fernandes, Nuno M. Silva, Carla Afonso, Marta D. Costa,

Eliska Musilová, Vincent Macaulay, Martin B. Richards, Viktor Černý, and Luísa Pereira.2012.The Expansion of mtDNA Haplogroup L3 within and out of Africa .Mol Biol Evol (2012) 29(3): 915-927 first published online November 16, 2011 doi:10.1093/molbev/msr245

Sun, Chang, Qing-Peng Kong, Malliya Gounder Palanichamy, Suraksha Agrawal, Hans Jurgen Bandelt, Yong-Gang Yao, Faisal Khan, Chun-Ling Zhu, Tapas Kumar Chaudhuri, and Ya-Ping Zhang.(2005). Molecular Biology and Evolution:

Wai-Ogusu,A.(1973). Was there a Sangoan industry in West Africa, West African Jour of Arcaheo,3:191-96.

Winters C. 2010. The Fulani are not from the Middle East. PNAS.
http://www.pnas.org/content/107/34/E132.extract

Winters C .2011. The Gibraltar Out of Africa Exit for Anatomically Modern Humans . WebmedCentral BIOLOGY 2(10):WMC002319 .
http://www.webmedcentral.com/article_view/2319

Winters, C.(2010b)Letter: The Fulani are not from the Middle East. PNAS.
http://www.pnas.org/content/107/34/E132.full

Winters, C. (2010c) The Kushite Spread of haplogroup R1*-M173 from Africa to Eurasia, Cur Res Jour of Bio Sci , 2(5), 294-299. http://maxwellsci.com/print/crjbs/v2-294-299.pdf

Wood,E.T., Stover,D.A., Ehret,C., Destro-Bisol,G., Spedini,G., McLeod, H., Louie,L., Bamshad,M., Strassmann,B.I., Soodyall,H., Hammer,M.F. (2005) Contrasting patterns of Y-chromosome and mtDNA variation in Africa:evidence for sex-biased demographic processes. Eur. J of Hum Genet, 13,867-876.

Yao, Y.G., Q.P. Kong, H.J. Bandelt, T. Kivisild and Y.P.Zhang, 2002. Phylogeographic differentiation of mitochondrial DNA in Han chinese. Am. J. Hum.Genet., 70: 635-651.

# Chapter Two: A Protocol to Evaluate Population Genetics Papers

There are tens of articles published each year in population genetics. These articles are must reading for anthropologist and molecular geneticists interested in migration and population genetics.

Using Bayesian statistics molecular geneticists make inferences about prehistoric demographic events, relating to various ethnic populations. To reconcile their genomic evidence with prehistoric and historical information some population geneticists use archaeological, linguistic and paleoanthropological data to corroborate their DNA findings.. The use of use archaeological, linguistic and paleoanthropological data to support molecular genetics is called archaeogenetics (Renfrew and Boyle, 2000).

Geneticists and anthropologists use Archaeogenetics to explain and discuss past population events. Archaeogenetics can be defined as the use of prehistoric and historical events to determined by archaeology, genetics and linguistics in concert with the DNA of various ethnic groups to infer the ethnic identity of ancient populations and/or the ancient migration of one population to another geographical location.

In population genetics the researcher usually uses the "wave of advance" model to explain demographic movements in the past. The "wave of advance" model was used to explain the spread of advantageous genes within a population( Ackland et al,2007; Renfrew, 2001).) . This theory was adapted to explain why an advantageous technology that may appear in one population spreads ( and or taken )to another population living in a different geographical area (Ackland et al, 2007).

Although archaeogenetics is the norm for many molecular geneticists, most researchers believe that Bayesian statistics alone, have sufficient power to demonstrate the valility of their research, and fail to corrobate the DNA data with corresponding archaeological, linguistic and paleoanthropological evidence.

Many people don't know how to evaluate population genetics articles, because they are expost facto research based on " statistical infererences" or the beliefs of the researcher supported by statistics. As a result, researchers can not judge the validity and reliability of the research. One must assume the research is correct based solely on the Bayesian statistical inferences—not the interactions between an independent variable and dependent variable(s).

In research there are two variables: variables that can be manipulated and variables that can not be manipulated.

A variable that can be manipulated is a variable that can be changed for example, your ability to perform a particular task can be influenced by the amount of training you receive in performing the task.

A variable that can not be manipulated can not be changed. For example, right now you are a particular age, it can not be manipulated. You are either Black or white, race can not change.

Research studies include a number of variables. Variables which can be manipulated or not manipulated

Independent Variable (IV) any variable used to control for individual differences (this variable usually not manipulated)

Dependent variable (DV) any outcome measure which is effected by the IV.    The effect of sex (IV) on reading achievement (DV).

Validity is testing the appropriateness, meaningfulness and usefulness of specific inferences made from test scores. In qualitative research the extent to which the research uses methods and procedures that ensure a high degree of research quality and rigor.

Internal Validity, we assume that whatever was manipulated produced a change in the dependent measure. IV insured by control of the extraneous variables: health, sex, race, SES, age, IQ, religion.

External Validity, provides the ability to generalize the findings. In other words the IV produced a change in DV.

In normal scientific research the researcher states a hypothesis and uses the scientific method to test his/her hypothesis. The validity and reliability of the piece of research is then determined by statistical significance tests focused on the interaction between the independent and dependent variable.

In the traditional evaluation of a piece of research literature you look at the researcher's hypothesis, results and statistical methods s/he used to determine the statistical significance of the research. This is not the case in population genetics research; in this research you are evaluating statistical inferences based on *the beliefs already held by the researcher* about a set of data, instead of testing a hypothesis.

As a result, the research contained in a population genetics article, reflects the views and beliefs already held by the researcher. Thusly, the statistical inferences will automatically support the views and beliefs held by that researcher; and any outliners that fail to support the researcher's beliefs may not be mentioned in the resulting research article/paper.

Here we will ask the question: "How do you evaluate population genetics research if it is expost facto research, that lacks an experimental design?" First, we will attempt to look at the doxa that may influence a geneticist's research and the constructs that should be considered when evaluating this knowledge base.

In reading any piece of research literature, wWe assume that any article or book written by an establishment member of the academe is reliable and valid. A piece of research full of valid scientific and/or historical truths--erudite scholarship and impeccable research based on the scientific method.

The scientific method is based on hypotheses testing. Hypotheses testing means that a researcher forms a hypothesis and test the hypothesis using a series of quantitative or qualitative statistical methods to determine the statistical significance of the hypothesis being tested. The scientific method is based on experimentation to test a hypothesis.

Population geneticists usually do not test hypotheses. They make inferences about data based on Bayesian statistical inferences. They do not use statistical methods to determine the statistical significance of a hypothesis, they use statistics to describe data being reviewed by the researcher based on the beliefs the researcher already holds about the data being reviewed.

Population genetics is a type of Expost facto research. Expost facto research design is a quasi-experimental type of study examining how an independent variable, present prior to the research study, affects a dependent variable.

Whereas the subjects in experimental research are randomly selected, the participants in Expost facto research, are not randomly selected or assigned. The genome of the research subjects is examined to determine the haplotypes and haplogroups carried by the participants in the study.

In population genetics research the researcher uses the Bayesian inference method of statistical inference. The Bayesian statistical method, is a subjective research design/method that provides a rational method of updating the researcher's beliefs.

Since, the results of a Bayesian statistical analysis are a series of beliefs based on statistical inferences, the results can not stand alone. This is due to the reality, that any results, reported by a researcher are only a series of inferences based on the researcher's belief about a phenomena backed up by a series statistical results. If the results are published without corresponding evidence from archaeology, anthropology, linguistics and or craniometrics the inferences are pure conjecture, because they reflect the attitudes already held by the researcher, confirmed by data selected by the researcher to support his or her beliefs.

There is a sociological basis behind how a researcher interprets data. Sociological research indicates that there are unconscious cognitive structures within each individual. Cognitive structures that hold the idealistic view of members of the academe that determine how they perceive "reality". These structures are called doxa (Berlinerblau 1999).

Commenting on these schema Berlinerblau (1999) noted that "These types of theories share the assumption that human beings know things that they do not even know that they know; that they "possess" knowledge about the world which exists in some sort of cognitive substrate, beyond the realm of discourse" (p.106). Wacquant (1995) says that doxa is " a realm of implicit and unstated beliefs".

Given the research suggesting that doxa exist, support the view that some researchers allow their hatred of multiculturalism, ethnic prejudice and racism to define their discourse, teaching and writing about themes relating to groups " other" , than their own cultural and ethnic group . Moreover, it suggest that when topics such as Eurasian and African haplogroups, Afrocentrism, African origins of the Dravidians and etc., is attacked by members of the academe, these academics are supported by the "establishment" without any reservation, or test of the validity of their claims. In fact, it appears that doxic assumptions relating to the validity of back migration of so-called Eurasian genes into Africa, recent African origin of Dravidians and Dravidian origin of the Indus Valley Civilization obviates critique of the academics that disparage these themes. Due to Doxa you can state a researcher's attitude toward a historical, genetic or anthropological concept and theorems without the statement being an ad hominem

To evaluate research literature a student should know the varied research methods. A student evaluating a piece of population genetics' literature must understand that the researcher is conducting an expost facto method of research that does not involve hypotheses testing. Given the nature of Bayesian inferences, you can not determine the validity and reliability of a piece of genetics research literature based on the statistical significance of the data. What you must do is look at the research article and ask yourself a series of questions regarding the article's validity and reliability.

To facilitate evaluation of genetics research literature I

have created a check list: Checklist used to analyze a Population Genetics Papers, to evaluate research articles.

To use the Checklist you would perform the following task. The Evaluator should read the article twice. The first reading of the article is brief.

Next make a close reading of the article. The close read should involve the Evaluator in underlining key details in the article, while making annotations of important points in the text.

During the second reading of the text the Evaluator will assess the research article using **the Checklist used to analyze a Population Genetics Papers**.Since the Bayesian statistics used for the study will support the inferences of the Researcher the answers for the majority of the questions on the checklist will be yes.

The key question in determining the validity of the research will be question 17. If the researcher only has Bayesian statistical inferences supporting the research study, the inferences made in the research article, may not be representative of actual past population events.

I will use the Checklist to evaluate a recent Population genetics article. The paper is Chaubey and Endicott (2015). As mentioned earlier Bayesian statitistics, since they are based on the author's belief, will just about always support the author's inference. Below are my responses to the article placed on the Checklist. The evaluation of this article revealed the following responses:

1-3 is yes

4. No

5. yes

6. yes

7. no

8. no

9. yes

10. yes

11. yes

12. yes

13. no

14. no. No discussion of Southeast Asian and mainland Indian archaeology.

15. yes

16. no

17. No

Because the answer to Question 17, was no, demands that we check the archaeology literature to determine if the Bayesian statistical inferences can find support from the craniometric, and archaeological record for SEA and India, or if the results and conclusion are based solely on the doxa of the researchers.

The claim that the Onge, a mainland Munda group only recently came to India circa 26kya. This would place them in India after the alledged settlement of India by the Aryans. They wrote: "of the Andaman-specific mtDNA lineage M31a1 around 26 kya, while the ages of the diversification within M32 and M31a1 are estimated to fall within the Holocene, using whole-genome data in a Bayesian statistical setting (Barik et al.2008). Because mtDNA divergence is anticipated to predate population divergence, collectively these estimates suggest that the Andamans were settled less than ~26 ka and that differentiation between the ancestors of the Onge and Great Andamanese commenced in the Terminal Pleistocene. Interestingly, this time frame is similar to the signal for population expansion found throughout ISEA (Guillot et al. this issue) and represents the time of topographic transition from the vast expanses of

Sundaland to the submerged Southeast Asian island chains of the Holocene. In conclusion, we find no support for the settlement of the Andaman Islands by a population descending from the initial out-of-Africa migration of humans, or their immediate descendants in South Asia. It is clear that, overall, the Onge are more closely related to Southeast Asians than they are to present-day South Asians.

The similarity in proportions of the Onge genomes, attributed to the Melanesian, Malaysian (Jehai and Kensui), and South Asian ancestral components, combined with evidence for genetic drift, suggests that these constituent parts were present prior to their isolation from other parts of Southeast Asia".

Although this is the opinion of Chaubey and Endicott (2015), the Onge and other Munda populations were in India long before the Aryans. C Winters (2010) argues that Thangaraj et al using coalescence time and archaeological evidence illustrated that the TRMCA for mtDNA R8 which is found among Munda speakers have the following dates : R8 (41.7 kya), R8a (15.4 kya) and R8b (27.7 kya)13. The dating for mtDNA R8 indicates that this haplogroup and R7 are probably autochthonus to India.

The mtDNA of Munda speakers also includes deep rooted haplogroups from macrohaplogroup M. In addition to mtDNA haplogroup M2, we also find M58, M31, M6a2 and M42 among Munda speakers.

The Munda y-chromosome is O2a (M95). Kumar reports a coalescent rate of 65kya for Indian M953. There is a clear distinction of Indian Munda and Southeast Asian

(SEA) Mon-Khmer speakers. The predominate SEA O clades are O3 and O1a. If SEA males had carried the y-chromosome O haplogroup to India there should be evidence of these clades among the Munda speakers—but they are nil8. On the otherhand, SEA males carry Indian y-chromosomes such as F,H, K2 (T) and etc8.

This indicates an early migration of Munda speakers to SEA. It suggest that Munda spread mtDNA R7 and y-chromosome haplogroup O to SEA.

Many Indians carry Munda haplogroups. The spread of Munda haplogroups are probably the result of conquest and intermarriage. The mythology of some Indian populations support this proposition. In other words, instead of the Munda originating in SEA, they probably migrated to the region from India.

Chaubey et al, based his conclusion on the research on Endicott et al (2006).Endicott et al (2006) argue that without comprehensive data from Myanmar it is not possible to identify whether the Andaman M31a1 arrived from India or if the Indian M31a2 came from South-East Asia. But either scenario casts serious doubts on the concept that the Andaman Islands were settled at the time of the migrations out of Africa carrying the current Eurasian mtDNA diversity".

Endicott et al (2006), admit that their conclusions should be preliminary because: "Without comprehensive data from Myanmar it is not possible to identify whether the Andaman M31a1 arrived from India or if the Indian M31a2 came from South-East Asia. But either scenario casts serious doubts on the concept that the Andaman Islands were settled at the time of the migrations out of Africa carrying the current Eurasian mtDNA diversity".

It is obvious that Endicott et al (2006) , could not answer this question because they did not know much about Southeast Asian history. If they knew the archaeology of Southeast Asia they would have been able to answer this question. They would have known that the Dravidians who carry M31a2 probably carry the haplogroup as a result of migration of Dravidian back to South India from Myanmar. Winters (2010), I explain0 that, many Dravidian speakers in India formerly lived in Southeast Asia.Formerly intimate relations existed between South Indians and Southeast Asian people (Kanakasahai, 1966). The Tamilian form of Saivism is known as Agamas, the esoteric and ritualistic parts of Agama is non-Vedic (not of Indo-European origin). Agama was also the Southeast Asian form of Hinduism (Winters,1985).

The Proto-Tamil speakers in Central Asia and China were called the Yakshas in Indian literature (Yuehchih by the Chinese) and Kosars (Kushana in Chinese literature). They were forced from China due to first the classical Mongoloids who founded Shang-Yin , then the

Zhou and succeeding mongoloid Chinese and Thai populations that invaded Indo-China. This forced the Proto-Tamil speaking Kosars and Yakshas to later invade southern India in search of a new homeland in addition to Southeast Asia (Winters, 2011). In Southeast Asia Dravidian speakers probably encountered Proto-Andamanese carrying M31 and M32 who may have been the original settlers of the area.

The archaeological, and genetic evidence indicate that Dravidian speakers lived in Southeast Asia (Kanakasabhai, 1966; Winters, 1985) . It indicates that the first civilizations in Southeast Asia were founded by Dravidian speakers (Kanakasabhai,1966).. The Khmer introduced various aspects of civilization in this region which precede the advent of the Thai speakers into this region. Upon their arrival in Indo-China ,the Thai-Vietnamese people conquered the blacks learned their culture and continued to perpetuate the same cultural traits (Winters,1985).Thusly, we see that both the Vietnamese and Thai peoples learned their culture, architecture, religion and writing from the Khmers and other Indo-African people.

While the Dravidians lived in Southeast they probably mated with the inhabitants related to the Andamanese (Winters,2011). This mating pattern probably led to M31a2 entering the Dravidian gene pool when the Kamboja settled in Sengal and South India

(Kanakasabhai, 1966).

In summary we can reject the research of Gyaneshwer Chaubey and Phillip Endicott, based on question 17 of the Checklist used to analyze a Population Genetics Papers, because it is unreliable and lacks validity because the researchers failed to study the archaeology and history of SEA. If they had, they would have known that Dravidian speakers formerly lived in SEA, until the advance of the Classical mongoloid people 2.5kya.

In summary, the validity and reliability of a piece of genetics research literature does not demand the Evaluator of a piece of literature to provide counter evidence all they need to do is evaluate the research using the checklist (see Appendix). If the answer to most of these questions is no, the research is unreliable and lacks any validity.

The key question on the checklist is question 17. To confirm the validity of the archaeological, craniometric and etc., data , the Evaluator should be knowledgeable about the archaeology of the area where the population movement has been inferred to have taken place. In this way you can determine if the Bayesian inferences correspond to the archaeological, craniometric, and linguistic data associated with the geographical area where the population movement is alleged to have occued .

The major problem with most genetics literature which invalidates the research dealing with ancient population movements is that it is not supported by the ancient DNA, archaeological and/ or craniometric data. This is why many of theories about the ancient populations of Europe and alledged back migrations are usually over turned once researchers examine the ancientDNA.

Reference:

Ackland G J , Markus Signitzer, Kevin Stratford,and Morrel H. Cohen.(2007).Cultural hitchhiking on the wave of advance of beneficial technologies PNAS 2007 104 (21) 8714-8719; published ahead of print May 16, 2007, doi:10.1073/pnas.0702469104. Retrieved 2/6/2015 at :
http://www.pnas.org/content/104/21/8714.full

Berlinerblau, J. (1999). Heresy in the University: The Black Athena Controversy and the Responsibilities of American Intellectuals .Rutgers University Press.

Chaubey, G and Phillip Endicott. (2013)The Andaman Islanders in a Regional Genetic Context: Reexamining the Evidence for an Early Peopling of the Archipelago from South Asia. Retrieved 3/6/2015 at:
http://digitalcommons.wayne.edu/cgi/viewcontent.cgi?article=2055&context=humbiol

Endicott P, Metspalu M, Stringer C, Macaulay V, Cooper A, et al. (2006) Multiplexed SNP Typing of Ancient DNA Clarifies the Origin of Andaman mtDNA Haplogroups amongst South Asian Tribal Populations. PLoS ONE 1(1): e81.
http://journals.plos.org/plosone/article?id=10.1371/journal.pone.0000081

Kanakasabhai,V.(1966). The Tamil Eighteen Hundred Years ago .

 Renfrew, C. (2001.From molecular genetics to archaeogenetics PNAS 2001 98 (9) 4830-4832; doi:10.1073/pnas.091084198.
http://www.pnas.org/content/98/9/4830.full


Winters, C. (1985).  "The Far Eastern Origin of the Dravidians", Journal of Tamil Studies, pp.66-92.

 Winters.C. (2010). Munda Speakers are the Oldest Population in India. The Internet Journal of Biological Anthropology. 2010 Volume 4 Number 2,
https://ispub.com/IJBA/4/2/5591

 Winters, C. 2011.  Comment : A back migration from Southeas Asia accounts for M31a2  in South India,
http://www.plosone.org/annotation/listThread.action?root=609

# The Phylogeography of Afro-Americans and Africans

# Appendix

### Checklist used to analyze a Population Genetics Papers

Answer the following questions relating to this research article below, or on a separate sheet of paper.

1.      What was the rationale for the study, that is, what led up to it? Yes on page____ , paragraph_____ _,lines_____ No_____

2.      Why do the authors believe that this problem is significant? Yes on page____ ,paragraph_____ _,lines_____ No_____

3.      What was the purpose of the study, that is , what did it intend to accomplish? Yes on page___ ,paragraph_____ _,lines_____ No_____

4.      What was the hypothesis of the study? Yes on page____ ,paragraph_____ _,lines_____ No_____

5.      What were the participants major characteristics? Yes on page___ ,paragraph_____ _,lines_____ No_____

6.      Does the review of literature indicate previous research in the area associated with the article? Yes on page___ ,paragraph_____ _,lines_____ No_____

7.      What type of study is reported in this article?Yes on page___ ,paragraph_____ ,lines_____
No_____

8.      Was the sample randomly selected?Yes on page___ ,paragraph_____ ,lines_____ No_____

9.      What was the instrument?Yes on page___ ,paragraph_____ ,lines_____ No_____

10.     What were the major steps involved in the treatment?Yes on page___ ,paragraph____ ,lines_____ No_____

11.     How were the variables tested? Yes on page___ ,paragraph_____ ,lines_____ No_____

12.   According to the author(s) how successful was the treatment?Yes on page___ ,paragraph____ ,lines_____ No_____

13.     What factors could equally account for the student tests results? Yes on page___ ,paragraph____ ,lines_____ No_____

14.     What problems, if any, do you detect in the study? Yes on page___ ,paragraph_____ ,lines_____
No_____

15.     Do the results of analysis agree with the authors objectives and expectations?Yes on page___ ,paragraph_____ ,lines_____ No_____

16.     What other interpretations could be made from the data?Yes on page____ ,paragraph____ _,lines_____ No_____

17.     Is there archaeological, craniometric and or linguistic evidence that supports the research findingsYes on page____,paragraph____ _,lines

# The Phylogeography of Afro-Americans and Africans

Dr. Clyde Winters

# Chapter Three: The African Origin of mtDNA Haplogroup M1

Abstract

Controversy surrounds the origin and expansion of the M1 haplogroup. Gonzalez et al (2007) and Olivieri et al. (2007) believe that the M1 macrohaplogroup originated in Asia and represents a backflow to Africa. The high frequency for M1 in Sub Saharan Africa instead of Asia and the Near East; the distribution of varied M1 haplogroups across Sub Saharan Africa; and the existence of an basal L3 motif in the Senegambia characterized by the DdeI site np 10394 and AluI site np 10397 in haplotype AF24 (DQ112852) which at the base of the M macrohaplogroup make a 'back migration' of M1 to Africa highly unlikely since this haplotype belongs to the rare haplogroup LOd which is not found in Asia, and probably spread to Senegal with amh using the Sangoan industry 80-50kya.

## Introduction

Controversy surrounds the origin and expansion of the M1 haplogroup. Gonzalez et al believe that the M1 macrogroup originated in Asia and represents a backflow to Africa**(Gonzalez et al, 2007)**. Other researchers believe that the M haplogroup originated in Africa **(Sun et al, 2006; Quintana-Murci et al ,1999)**. Quitana-Murci et al has suggested that M1 probably originated in Ethiopia prior to the out of Africa migration 60kya ( sixty-thousand years ago**). (Quintana-Murci et al, 1999)**.

The M1 haplogroup is a member of the M macrohaplogroup. M1 is a sister haplogroup to Haplogroup D, one of the major Asian subgroups in Macrohaplogroup M. The M,N, and R macrogroups are found throughout East and South and Southeast Asia, the Andaman Islands and Africa **(Ingman et al, 2000, 2003;Macaulay et al, 2005; Tanaka et al, 2004)** .

The M haplogroup was probably part of the original out of Africa event around

60,000 ya **( Kivisild et al, 2004; Macaulay et al, 2005; Rando et al, 1998; Tanaka et al, 2004; Sun et al, 2006)**, but there is controversy over wether haplogroup M1 originated in Africa, or later in Asia.

The transitions 489T>C, 10400C>T, and 15043G>A define Haplogroup M, though in the early studies membership in Haplogroup M was usually determined from the results of an RFLP test—an AluI site of np

10397 ( an indicator of 10400T). All members of

haplogroup M also have other well known differences

From CRS, namely 10398G and 10873C, which are the

ancestral states, compared to the mutation 10398G>A

and 10873C>T that occurred in Haplogroups N and R on

the line to CRS, and the HVR1 mutation 16223T>C ,

which also occurred in haplogroup R, also on the line to

CRS. Haplogroup M originated from an African

Haplogroup L3 background.

**Materials and Method**

We analyzed the mtDNA sequences of the M

haplogroup from Africa, Asia, India, Southeast Asia and

Oceania from the literatures (Gonzalez et al, 2006,2007;

Ingman et al, 2000, 2003; Kivisild et al, 1999, 2004;

Macaulay et al, 2005; Rajkumar et al, 2005; Sun et al,

2005; Tambeto et al, 2000; Tanaka et al, 2004). This meta-analysis of mtDNA literatures allowed us to critically look at the distribution of M1 alleles across the Macrohalogroup M.

**Results**

The M1 macrohaplogroup is found throughout Africa and Asia. But the basal M1 lineage has not been found outside Africa **( Kivisild et al, 2004; Rajkumar et al, 2005;Sun et al, 2005).**

The Haplogroup M1 branch is defined by several mutations, including 195T>C, 16129G>A, 16249T>C and 16311 T>C, in the control region, and 6446G>A,6680T>C,12403A>C and 14110T>C **(Sun et al, 2005).** The RFLP of M1, considered diagnostic in many early studies, is by MnII site loss at 12402 (an indicator of 12403T).

Gonzalez et al **(2007)**, believes that M1 originated in Asia and the most ancient M1 sublineage (called M1c by Gonzalez et al), originated in Northwest Africa, though he reports that the highest frequency of M1 is found in Sub-Saharan Africa especially East Africa.

Haplotypes with HVSI transitions defining 16129-16223-16249-16278-16311-16362; and 16129-16223-16234-16249-16211-16362 have been found in Thailand and among the Han Chinese **(Fucharoen et al, 2001; Yao et al, 2002)** and these were originally thought to be members of Haplogroup M1. However, on the basis of currently available FGS sequences, carriers of these markers have been found to be in the D4a branch of Haplogroup D, the most widespread branch of M1 in East Asia **(Fucharoen et al, 2001; Yao et al, 2002)**. The transitions 16129,16189,16249 and 16311 are known to be recurrent in various branches of Haplogroup M, especially M1 and D4.

Earlier researchers failed to find M1 among the M lineages in India **(Rajkumar et al, 2005; Olivieri et al, 2006)**. Gonzalez et al (2007) also stated that the M1 HVSI diagnostic motif has not been found among Indian M haplogroups .

Gonzalez et al **(2007)** divided M1 primarily into to subgroups which he called M1a and M1c, which are now named M1a1 positions 3705,12346 and M1b position 13111. M1a1 may be identified when only the HVS1 data is available by its characteristic 16359C. The transitions 16260 and 16182 occur in subgroups of for M1a1, so they too may be used to identify a M1a1 sample. M1a1 lineages are frequently found in Ethiopia and East Africa. The M1a2 position 15884 (called M1b by Gonzalez et al) branch is identified by an HVS1 transitions at 16185, and a deletion at 16190 deletion, is common in West Africa and Jordan. Gonzalez et al (2007) notes that the M1a2 (his M1c) clade is found in Northwest and West Africa.

To estimate the coalescence age of haplogroup M1 Gomzalez et al (2007) analyzed 13 complete sequences of haplogroup M1.

Gonzalez et al **(2007)** claims that the M1c lineage is the oldest M1 subclade based on the coalescence age estimation of the M1 subgroup: M1a (16756 +-5997), M1b (10155 +-3590) and M1c (19040+-4916). This makes M1a and M1b the youngest clades.

The available sample for M1c was complete sequences from individuals found in Jordan, Senegal, and Spain. The small data set make a precise estimation of the errors in the data uncertain.

The limited sample for M1c makes it difficult to

effectively quantify the estimation error for the data, since error increases from level to level in models possessing a hierarchical structure.

The small sample size makes the confidence intervals overlap. This calls into question the conclusions of Gonzalez et al (2007) despite the differing levels of hierarchy.

If the sample used by Gonzalez et al (2007) had been larger we might expect the researchers to have paid close attention to the estimated value of the variance in the data sets. Given the extremely small size of the data set, the researchers probably has too much confidence in the predicted ages for the M1 subsets, because the sample was too small to allow the estimation errors to propagate as the data was analyzed. The failure to effectively estimate uncertainty in the limited data set probably led to estimation errors in the predicted ages for the M1 subclades, which inflated the age of M1c in relation to the other M1 subsets.

**Table 1:Continental Population frequencies, Means &**

**Standard Deviations for Haplogroups M1 and M1a**

**based on Gonzalez et al.**

| Haplogroup | M1 | M1a | | |
|---|---|---|---|---|
| Countinent | Percent | Percent | Mean | S D |
| Europe | 17.9 | 11.0 | 14.5 | 3.45 |
| North Africa | 13.7 | 2.0 | 7.85 | 5.85 |
| Africa | 55.4 | 68.0 | 61.7 | 6.3 |
| Asia | 13.0 | 19.0 | 16 | 2.85 |
| | | | | |
| | | | | |

In addition to the evidence of the coalescence age

estimation in support of the antiquity of M1c, Gonzalez

et al **(2007)** believe the presence of M1c among

Jordanians is an important indicator for the ancient

origin of this clade. The evidence of M1c in Jordan, does not really add to the hypothesis that M1c is the oldest clade because the presence of this clade in the Middle East can be explained by the thousands of West Africans who have taken the hajj to Mecca, and remained in the Middle East, instead of returning to West Africa.

The Valencia sample can also be explained by the history of Islam. There is a direct link between Senegal and Tariq ibn Ziyad's invasion of Spain in 711. This link comes from the fact that many of the followers of Tarik came from the *ribats* or 'religious schools' he had established in northern Senegal. Troops from these ribats formed the backbone of Tarik's army. These African Muslims ruled much of Spain until 1492. Since M1c is presently found in Senegal, the carrier of M1c reported by Gonzalez et al (2007) in Valencia may be a descendent of these African 'Moors' that ruled Spain for over 700 years.

The results published by Gonzalez et al (2007) fail to support his conclusion. In Table 1, we see the geographical ancestry of the groups used in this study. The percentage of individuals carrying this haplogroup in this study was between 13.0% and 55.4% for M1 and 11.0% and 68.0 % for M1a1.

The distribution of continental populations carrying the M1 haplogroup favors Africa as the place of origin instead of Asia for both haplogroup M1 (55.4%) and haplogroup M1a1 (68.0%). The population distributions for both M1 and M1a make it clear that the most varied M1 subhaplogroups appear across Sub-Saharan Africa, not Asia or North Africa.

## Figure 1: Continental Distribution of M1

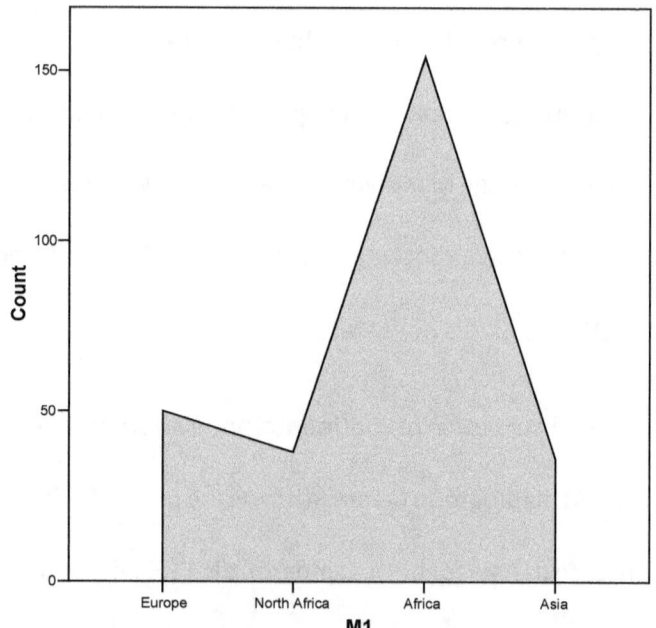

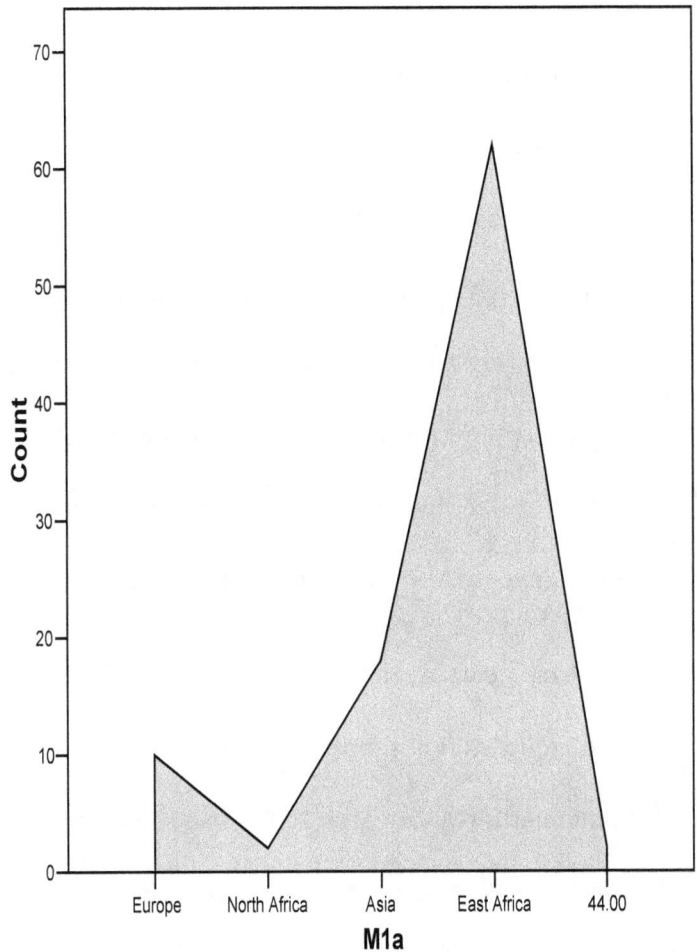

**Figure 2: Continental Distribution of M1a**

Gonzalez et al **(2007)** claim that their research is supported by the research of Olivieri et al **(2007)**. Anna Olivieri et al **(2007)** claims that M1 originated in Asia and the M1 haplogroup represents a back-migration into Ethiopia. Olivieri et al (2007) based this conclusion on: 1) the absence of any distinguishing M1 root mutations in Asian M haplogroups; 2) the presence of M1only in East Africa and North Africa; and 3) the lack of any Asian specific clades within M1. These conclusions by Olivieri et al (2007) are incongruent with the evidence of M1 transitions in Asian M clades.

The molecular evidence makes it clear that haplogroup M1 is not confined solely to Ethiopia as maintained by Olivieri et al (2007). This haplogroup along with HGs N and M*, are also found in Tanzania, Uganda, Egypt and the Senegambian region **(Gonzalez et al, 2006; Gonder et al, 2006; Winters, 2007)**.

In Tanzania the predominate M1 clades are M1, M1a1 and M1a5. In Senegal the predominate M1 lineage is M1c1.

In addition to haplogroups M1, M* and N in Sub-Saharan Africa we also find among the Senegambians hapotype AF24 (DQ112852), which is delineated by a DdeI site at 10394 and AluI site of np 10397. This haplotype is a branch of the African subhaplogroup L0d. It clear from the molecular evidence that the M1, M and N haplogroups are found not only in Northeast Africa, but across Africa from East to West **(Winters, 2007)**.

Haplotype AF-24 is an ancient African haplotype. AF-24 is aligned to the Asian M macrohaplogroup.

The Senegalese haplotype AF-24 belongs to the rare ancient mtDNA haplogroups L0d. The L0d haplogroup is limited only to West Africa (Rosa et al, 2004), East

Africa and South Africa (Gonder et al, 2006).

Haplogroup LOd is found at the root of human mtDNA. Gonder et al (2006) maintains that LOd is "the most basal branch of the gene tree". The TMRCA for LOd is 106kya. This makes haplotype AF-24 much older than L3a.

Since the TMRCA of LOd dates to 106kya Anatomically modern humans (amh) had plenty of time to take this haplogroup to Senegal. In West Africa the presence of amh date to the Upper Palaeolithic (Giresse, 2008).

The earliest evidence of human activity in West Africa is typified by the Sangoan industry (Phillipson, 2005). The amh associated with the Sangoan culture may have deposited Hg LOd and haplotype AF-24 in Senegal thousands of years before the exit of amh from Africa.

Anatomically modern humans arrived in Senegal during the Sangoan period. Sangoan artifacts spread from East Africa to West Africa between 100-80kya. In Senegal Sangoan material has been found near Cap Manuel (Giresse, 2008), Gambia River in Senegal (Davies, 1967; Wai-Ogussu, 1973); and Cap Vert (Phillipson, 2005).

Olivieri et al (2006) provide a detailed discussion of the M1 macrohaplogroup. The distribution of the M1 superhaplogroup is outlined in Table 3. Here we note that as in the Gonzalez et al. study, the widest distribution of M1 clades appear in Sub Saharan Africa, not the Near East or the Mediterranean region (see Figure 3).

## Table 2: Distribution of M1 Subhaplogroups, taken from Oliveri et al (2006)

| Haplogroup | M1a | Mla l | M1al a | M1a1 b | M1alb 1 | M1a1 c | M1al d | M1al e |
|---|---|---|---|---|---|---|---|---|
| Country | | | | | | | | |
| Med N | - | - | 4 | 5 | 30 | - | - | 45 |
| Med S | - | - | 7 | - | - | - | - | 21 |
| Near East | - | 20 | 4 | - | - | - | - | - |
| Egypt | - | - | - | - | - | - | - | - |
| East Africa | 74 | 33 | - | - | - | 21 | 54 | - |

## Figure 3: Olivieri et al. Normative M1 Groups by Country

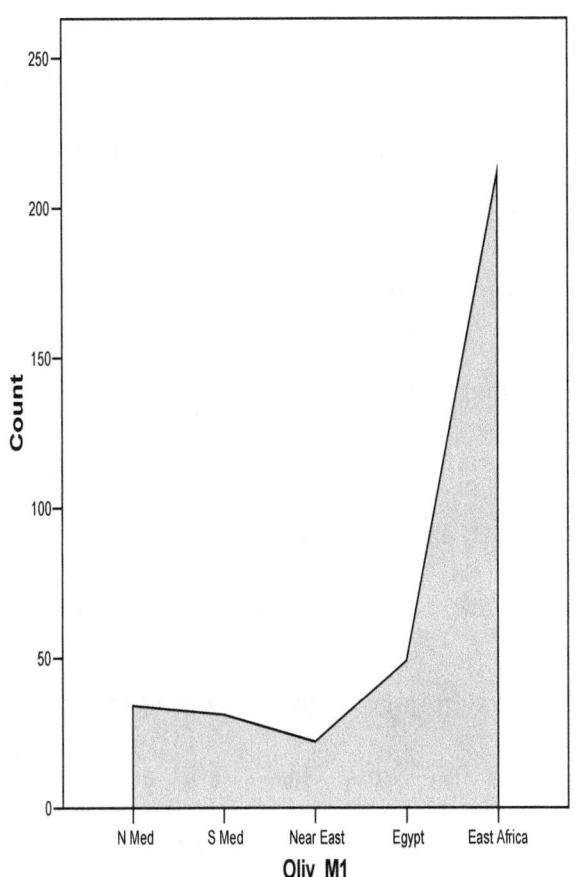

Haplogroup M1 is found throughout Africa. The diversity haplogroup M1 on in Sub-Saharan Africa makes it clear that it could not be the result of a back migration.

The predominate language spoken by carriers of M1 and M* in Africa outside of Ethiopia are Niger Congo speakers. The people in Ethiopia and Egypt that carry M1 hg speak Afro-Asiatic languages.

The fact that haplogroup M1 is found among Niger-Congo speakers and Afro-Asiatic speakers is telling given the fact that the both the Niger-Congo speakers and Afro-Asiatic speakers are associated with the Nubia.

The major cultural group in Nubia during the Neolithic was the C-Group. The C-Group people of Nubia are associated with the rise of millet cultivation in Africa, and spread of this crop and the red-and-black ceramic style from Africa to India (Singh 1982;

Winters, 2008b)

Welmers (1971) explained that the Niger-Congo

homeland was in the vicinity of the upper Nile valley

(p.119). He believes that the Westward migration out of

Nubia into West Africa began 5000 years ago.

In support of this theory he discusses the dogs of

the Niger-Congo speakers. This is the unique barkless

Basenji dogs which live in the Sudan and Uganda today,

but were formerly recorded on Egyptian monuments

(Wlemers, p.119). According to Welmers the Basanji, is

related to the Liberian Basenji breed of the Kpelle and

Loma people of Liberia. Welmers believes that the

Mande took these dogs with them on their migration

westward. The Kpelle and Loma speak Mande

languages.

Welmers (1971) proposed that the Niger-Congo speakers remained intact until 5000 years ago. This view is supported by linguistic and genetics evidence. The linguistic evidence makes it clear that the Nilo-Saharan and Niger-Congo languages are related. The genetic evidence indicates that Nilo-Saharan and Niger-Congo speakers carry the M3b*-M35 gene, an indicator for the earlier presence of speakers of this language in an original Nile Valley homeland.

# Map 1 :Geographical Distribution of M1 Haplogroup in Africa.

The distribution of M1 in Africa is widespread. We see evidence of this haplogroup from East to West and even down into South Africa. M1 is found among many African groups especially the Niger-Congo speakers. This is very interesting because most researchers have established the origin of this linguistic family in Nubia. From here the speakers of this language family migrated in West Africa and East Africa. The numbers indicate the carriers of M1 in the various African nations.

Linguistic research make it clear that there is a close relationship between the Niger-Congo Superlanguage family and the Nilo-Saharan languages spoken in the Sudan. Heine and Nurse (2000), discuss the Nilo-Saharan connection. They note that when Westerman (1911) described African languages he used lexical evidence to include the Nilo-Saharan and Niger-Congo languages into a Superfamily he called "Sudanic" (p.16). Using Morphological and lexical similarities Gregerson (1972) indicated that these languages belonged to a macrophylum he named " Kongo-Saharan" (p.16). Research by Blench (1995) reached the same conclusion, and he named this Superfamily: "Niger-Saharan".

There was a close relationship between the the Niger-Congo speakers in Nubia and the Ethiopians. This view is supported by the archaeological evidence that support a close relationship between the Ethiopians and Nubians. For example, according to Fattovich (2008) the pottery from Tihama Cultural Complex and other Ethiopian sites shows similarities to the Kerma and C-Group pottery. Given this connection between Ethiopian civilizations and civilizations in Nubia, may explain the presence of the M1 haplogroup among people in East and West Africa which formerly lived in intimate contact.

**Map M Haplogroups in Africa.** The map makes it clear that M haplogroups so far found in Africa vary. They not only vary but illustrate the expansion of the Niger-Congo and Nilo-Saharan speakers. Both groups originally lived in Nubia and may have been part of the C-Group people of Nubia.

The earliest civilization in Southwest Arabia date back to the 2nd Millenium BCE. This culture is called the Tihama culture which originated in Africa (Fattovich, 2008). Keall (2008) believes that these people may have had a common ancestry and shared a common culture.

The Tihama civilization probably originated in Nubia. It is characterized by the cheesecake or pillbox burial monuments which extend from Dhofar in Nubia, the Gara mountains to Adulis on the Gulf of Zula, to Hadramaut, Qataban, Ausan, Adenm, Asir, the Main area and Tihama.

Shared culture of the C-Group people (who probably included many Niger-Congo speakers would explain the affinity between the earliest Ethio-Semitic culture

:Tihama and the C-Group. At Tihama and other sites in Arabia we find pottery related to the C-Group people of Nubia (Keall, 2000;2008; Fattovish, 2008; Giumlia-Mair, 2002)The archaeological evidence indicates that C-Group people expanded from Nubia to Mesopotamia and the Indus Valley.

It appears that whereas the Egyptians preferred the cultivation of wheat, many ancient C-Group people were agro-pastoral people who cultivated Millet/Sorghum and rasied cattle. Millet was the main crop of the Dravidian speakers of India and people of the Indus Valley (Possehl, 1986; Winters,2008a).

Numerous linguists and anthropologists claim that the Dravidians originated in Africa. For example, B.B. Lal (1963) , a leading Indian archaeologist made it clear that he saw a relationship between the C-Group people of Nubia and Dravidian speakers. In relation to the anthropological evidence Aravanan (1976,1979,1980),

Homburger (1930,1951,1964), Sergent (1992), Sastri

(1966), Lahovary, Upadhyaya (1976,1979) claim that the

Niger-Congo and Dravidian languages are genetically

related.

## Discussion

There is considerable evidence that M1 is found in

Asia. Researchers have found the M1 haplogroup in the

Caucasus **( Bermisheva et al, 2004; Tambets et al,**

**2000)**, Central Asia, and East Asia **(Comas et al, 1998;**

**Fucharoen et al, 2001)**. In addition, the Russian

haplotype 16183c-16189 ,16249, 16311 match the M1

HVSI sequence **(Malyarchuk et al, 2004)**.

Oliviera et al **(2007)** claim that East African M1 root

mutations are absent in Eurasian M sister clades is not

supported by the evidence. For example researchers

have found that the Tanzanian M1 haplogroup cluster

with people from Oceania **(Gonder et al, 2006)**. And, as

mentioned earlier the M1 mutations 16129, 16189,

16249 and 16311 are found in many southeast and East

Asian haplogroups **(Fucharoen et al, 2001; Yao et al,**

**2002)**. In addition,

Roychoudhury et al (2001) noted defining nucleolides

shared by East African M1, and Indian M haplogroups

include HG M4 at 16311 ; HG M5 at 16,129; and HG

M34 at 16,249 ; and Sun et al (2005) found that the

most frequent transitions in Indian M haplogroups

were 16129, 16311 and 16189 **(Sun et al, 2005)**. It is

interesting to note that whereas 489c is found in

Eastern African and Indian M mtDNA analyzed, it was

not found in the non M haplogroup controls.

It is also not true that HG M1 is absent in India.

Gonzalez et al (2007) report the presence of one

individual carrying hg M1 in the Appendix of their study. Kivisild et al **(1999)** noted that 26 of the subjects in his study belonged to the M1 haplogroup. Kivisild et al (1999) reported subcluster M1 was found mainly in Kerala and Karnataka high caste individuals **(Kivisild et al, 1999)**.

Kivisild et al (1999) made it clear that each Indian M lineage has its own unique star features. Kivisild et al **( 1999)** found 5 major haplogroup M subclusters in India :Haplogroup M1 transitions at 16129,16189 and 16311 (were reported by Kivisild et al 1999 in Figure 3), Haplogroup M2 447G,1780,11083,15670, 16274 and 16319 transitionsHaplogroup M3, Haplogroup M4 position 16311, and Haplogroup M5 positions 1888 and 16129.

Chaubey et al (2006) claims that the haplogroup Kivisild et al (1999) identified as M1, is now HG M3.

But this can not be supported by the reading of Kivisild et al (1999) phylogeny of Indian M haplogroups outlined in Figure 3, makes this unlikely. A cursory examination of Figure 3 (Kivislid et al, 1999), makes it clear that these researchers found different transitions for the Indian M1 and M3 subclusters as reported in their paper.

Olivieri et al **(2007)** argue that M1 was probably spread from the Levant back into Africa by the Aurignacian culture. The craniofacial and molecular evidence does not support this conclusion **(Winters, 2007)**.

There have been numerous Sub-Saharan skeletons found in the Levant and Europe **(Barral, 1963; Diop, 1974,1991; Brace, 2006;Boule,1957, Verneaux, 1926)**. Moroever, the identification of Sub-Saharan craniometric features **(Ehret, 1979; Haak et al , 2005;**

**Holliday,2000; Wendorf,1968)** and N haplogroup

among ancient skeletons found at sites in Europe

formerly occupied by Cro-Magnon people **(Haak et al,**

**2005)** , suggest that the ancient Sub-Saharans in the

Levant and Europe already possessed the N (and M)

haplogroup(s) when they arrived in these areas from

Africa.

In conclusion, we must reject the contention of

Gonzalez et al. **(2007)** and Olivieri et al. **(2007)** that M1

originated in Asia because of the possible Senegalese

origin of the M1c subclade; the absence of the AF-24

haplotype of haplogroup LOd; and African origin of the

Dravidian speakers of India .

Gonzalez et al (2007) reported that the carriers of

the M1c subset were from Jordan, Senegal and

Valencia. It was revealed above that 1) many of the

Muslim troops in Tarik's army that conquered Spain in

711 AD, came from Senegal; and 2) many West Africans after taking the Hajj, visited Jerusalem and settled in the Middle East. Even if we eliminate the Jordan sample, the evidence from Valencia and Senegal gives a 67% probability that M1c originated in Senegal, not Asia or North Africa.

It must also be rejected because of the craniometric and genetic evidence that the Dravidian speakers who carry the Indian M haplogroups originated in Africa and spread to India within the past 4500 years **(Lal, 1963; Winters, 2007, 2008b)**. Although Gonzalez et al **(2007)** attempts to use Olivieri et al **(2007)** to support their research, the latter paper fails to offer it support. The research indicates that contrary to Olivieri et al. **(2007)** view, M1 transitions are found in many Asian M haplogroups **(Fucharoen et al, 2001; Yao et al, 2002)** , the distribution of the M1 haplogroup across Sub

Saharan Africa (See: Map 1)

and the greatest diversity theory support an African

origin of M1, rather than an Asian origin because most

diversified M1 lineages are found in Sub Saharan Africa,

not Europe or Asia and there are other M haplogroups

in Africa that can not be accounted for as back

migrations from Asia to Africa (See Map 2).

Gonder et al (2006) argues that the TMRCA of mtDNA

L3(M,N) and their derivatives is around 94.3kya. It was

not until 60kya that the TMRCA of non-African L3(M,N)

exited Africa. This was 30,000 years after the rise of L3

and L0d and predicts a significant period of time for

anatomically modern humans (amh) living in Africa to

spread L3(M) haplogroups across the continent. The

existence of the basal L3a(M) motif and the L0d

haplotype AF-24 among Senegalese supports this view.

Gonder et al (2006) claimed that LOd is exclusive to the southern African Khoisan (SAK) population. The presence of the ancient AF-24 haplotype among the Senegalese, that is absent in other parts of Africa, suggest a long-term population in the Senegambia that preserved this rare haplotype—that originated early in the history of amh.

Moreover, the existence of the L3a-M motif in the Senegambia characterized by the Ddel site np 10394 and Alul site np 10397 in haplotype AF24 (DQ112852) make a 'back migration of M1 to Africa highly unlikely, since this haplotype is associated with LOd . The first amh to reach Senegal belonged to the Sangoan culture which spread from East Africa to West Africa probably between 100-80kya.

The reality that AF-24 is a haplotype of haplogroup LOd makes it clear that this haplotype is not only an ancient human genome, it is also evidence that AF-24

probably did not originate in Asia, since it was found

among the Senegalese, and reflects an early migration

from East Africa to West Africa. The presence of the

nucleotides characteristic of macrohaplogroup M in

Africa and the reality that M1 does not descend from an

Asian M macrohaplogroup because of the absence of

AF24 in Asia **(Sun et al, 2005)** suggest that expansion of

M1 was probably from Africa to Eurasia. The existence

of haplotype AF-24 and the LOd lineage in East and

West Africa also implies the probable existence of the

Proto-M1 lineage in Africa, not Eurasia.

References:

Aravanan, K P .(1976). "Physical and cultural

similarities between Dravidians and Africans", **Journal**

**of Tamil Studies** 10: 23-27.

Aravanan, K P . (1979) **Dravidians and Africans** ,

Madras.

Aravanan,K.P. (1980) Notable negroid elements in Dravidian India, **Journal of Tamil Studies**, pp.20-45.

Barral,L. & Charles,R.P: **Nouvelles donnees anthropometriques et precision sue les affinities systematiques des negroides de Grimaldi,** Bulletin du Musee d'anthropologie prehistorique de Monaco 1963, No.10:123-139.

Behar DM, Villems R, Soodyall H, Blue-Smith J, Pereira L, et al. The dawn of human matrilineal diversity. *Am J Hum Genet.* 2008;82:1130–1140. [PubMed]

Bermisheva MA, Kutuev IA, Korshunova TY, Dubova NA, Villems R, Khusnutdinova E: **Phylogeographic analysis of mitochondrial DNA in the Nogays: A strong mixture of maternal lineages from eastern and western Eurasia.** *Molec Biol 2004* , **38**:516-523.

Blench, R. 2006. *Archaeology, Language, and the African Past* New York: Altamira Press

Boule, M., HV Vallois : **Fossil Man** . Dryden Press New York , 1957.

Brace, C.L. , Noriko Seguchi, Conrad B. Quintyn, Sherry C. Fox, A. Russell Nelson, Sotiris K. Manolis, ** and Pan Qifeng: The questionable contribution of the Neolithic and the Bronze Age to European craniofacial form. **Proc Natl Acad Sci U S A**. 2006 January 3; 103(1): 242–247.

Chaubey G, Metspalu M, Villems R, Kivisild V. (2007). Reply to Winters. **BioEssays** 29(5):499.

Chen YS, Olckers A, Schurr TG, Kogelnik AM, Huroponen K, Wallace DC. (2000). mtDNA variation in the South African Kung and Khwe — and Their genetic relationships to other African populations. **Am J Hum**

**Genet**, 66(4): 1362-1383.

Comas D, Calafell F, Mateu E, Pérez-Lezaun A, Bosch E, Martínez-Arias R, Clarimon J, Facchini F, Fiori G, Luiselli D, Pettener D, Bertranpetit J: **Trading genes along the silk road: mtDNA sequences and the origin of Central Asian populations.** *Am J Hum Genet* 1998, **63**:1824-1838.

Davies,O. (1967). <u>West Africa before the Europeans</u>. London.

Diop,A: **The African Origin of Civilization**. Lawrence Hill Books,1974 .

Diop, A: **Civilization or Barbarism**. Lawrence Hill Books, 1991.

Ehret,C. :**On the antiquity of agriculture in Ethiopia**, Jour. of African History 1979, 20, p.161.

Fattovich, R. (2008). **The development of urbanism in the Northern Horn of Africa in ancient and Medieval Times**. Retrieved 2/19/2008

http://www.arkeologi.uu.se/afr/projects/BOOK/fattowich.pdf

Fucharoen G, Fucharoen S, Horai S: **Mitochondrial DNA polymorphism in Thailand**. *J Hum Genet* 2001, **46**:115-125.

Gonder MK, Mortensen HM, Reed FA, de Sousa A, Tishkoff SA: **Whole mtDNA Genome Sequence Analysis of Ancient African Lineages**. Mol Biol Evol. 2006, Dec 28.

*González, A. M., V. M. Cabrera, J. M. Larruga, A. Tounkara, G. Noumsi, B. N. Thomas and J. M. Moulds:* **Mitochondrial DNA Variation in Mauritania and Mali and their Genetic Relationship to Other Western**

**Africa Populations**. Annals of Human Genetics 2006,

**70,5.** http://www.blackwell-

ynergy.com/doi/abs/10.1111/j.1469-

1809.2006.00259.x?cookieSet=1&journalCode=ahg

Gonzalez, A. Jose M Larruga, Khaled K Abu-Amero,

Yufei Shi, Jose Pestano and Vicente M Cabrera.

(2007)**Mitochondrial lineage M1 traces an early human**

**backflow to Africa**, BMC Genomics, 8:223

doi:10.1186/1471-2164-8-223.

Giumlia-Mair, A., Keall, E. J., Shugar, A. and Stock, S.

(2002) .Investigation of a Copper-based Hoard from the

Megalithic Site of al-Midamman, Yemen: an

Interdisciplinary Approach=, Journal of Archaeological

Science 29, 195-209.

Giresse, P. (2008). Tropical and sub-Tropical West

Africa—marine and Continental changes during the late

Quaternary. Volume 10. Elsevier Science.

Haak, W. Peter Forster, Barbara Bramanti, Shuichi Matsumura, Guido Brandt, Marc Tänzer, Richard Villems, Colin Renfrew, Detlef Gronenborn, Kurt Werner Alt, Joachim Burger: **Ancient DNA from the First European Farmers in 7500-Year-Old Neolithic Sites.** Science 11 November 2005: 310(5750) 1016 – 1018.

Heine and Nurse (Eds.). (2000). **African languages: An introduction**, Cambridge University Press.

Holiday, T: **Evolution at the Crossroads:Modern Human Emergence in Western Asia.** American Anthropologist 2000, 102(1).

Homburger,L.1930. "Dialectes coptes et Manding". **BULL. SOC. LING.**,Tome 30.

_____. 1951. "Le. Telugu et Mende dialects".
JOURNAL DE LA SOC. AFR.

_____. 1964. "Sibilants en Sindo-Africaine".
JOURNAL DE LA SOC. AFR.

Ingman M, Kaessmann H, Pääbo S, Gyllensten U:
Mitochondrial genome variation and the origin of
modern humans. Nature 2000, **408**:708-713.

Ingman M, Gyllensten U: **Mitochondrial genome
variation and evolutionary history of Australian and
New Guinean aborigines**. Genome Res 2003, **13**:1600-
1606.

Keall, E. J. (2000) .Changing Settlement along the Red
Sea Coast of Yemen in the Bronze Age, First
International Congress on the Archaeology of the
Ancient Near East (Rome May 18-23, 1998),
Proceedings, (Matthiae, P., Enea, A., Peyronel, L. and

Pinnock, F., eds), 719-31, Rome.

Keall, **Dr. Edward J. Contact across the Red Sea (between Arabia and Africa) in the 2nd millennium BC: circumstantial evidence from the archaeological site of al-Midamman, Tihama coast of Yemen, and Dahlak Kabir Island, Eritrea** . Retrieved 2/20/08 at:

http://72.14.205.104/search?q=cache:SJPE_UY0VWUJ:www.dur.ac.uk/resources/mlac/arabic/RSPIabstracts02.pdf+keall,+Contact+across+the+Red+Sea+(between+Arabia+and+Africa)+in+the+2nd&hl=en&ct=clnk&cd=1&gl=us

Kivisild, Toomas, Katrin Kaldman, Mait Metspalu, Juriparik, Surinder Papiha: **The Place of the Indian mtDNA Variants in the Global Network of Maternal Lineages and the Peopling of the Old World.** In Genomic Diversity, (Ed.) R. Papiha Deka (pp.135-152). S.S. Kluwer/Plenum Publishers 1999.

.http://evolutsioon.ut.ee/publications/Kivisild1999b.pdf

Kivisild, T. Maere Reidla, Ene Metspalu,Alexandra Rosa, Antonio Brehm,2 Erwan Pennarun, Jüri Parik, Tarekegn Geberhiwot, Esien Usanga, and Richard Villems: **Ethiopian Mitochondrial DNA Heritage: Tracking Gene Flow Across and Around the Gate of Tears**. Am J Hum Genet. 2004 November; 75(5): 752–770.

Lal, BB. 1963. "The Only Asian Expedition in threatened Nubia:Work by an India Mission at Afyeh and Tumas". **THE ILLUSTRATED TIMES,** 20 April.

Malyarchuk B, Derenko M, Grzybowski T, Lunkina A, Czarny J, Rychkow S, Morozova I, Denisova G, Miscicka-Sliwka D: **Differentiation of mitochondrial DNA and Y chromosomes in Russian populations**. *Hum Biol* 2004, **76**:877-900.

Macaulay V, Hill C, Achilli A et al. (21 co-authors): **Single, rapid coastal settlement of Asia revealed by analysis of complete mitochondrial genomes**. Science 2005, **308**:1034–1036.

Olivieri A, Achilli A, Pala M, Battaglia V, Fornarino S Al-Zahery N, Scozzari R, Cruciani F, Behar DM, Dugoujon JM, Coudray C, Santachiara-Benerecetti AS, Semino O, Bandelt HJ, Torroni A: **The mtDNA legacy of the Levantine early Upper Palaeolithic in Africa**. *Science 2006* , **314**:1767-1770.

Phillipson, D.W.(2005). African Archaeology. Cambrige.

Possehl, G L. (1986). African Millets in South Asian Prehistory. In J. Jacobson (Ed.), **Studies in the archaeology of India and Pakistan** (pp.237-256). New Delhi: Oxford and IBH.

Quintana-Murci L, Semino O, Bandelt H-J, Passarino G, McElreavey K, Santachiara-Benerecetti AS. (1999) **Genetic evidence of an early exit of Homo sapiens sapiens from Africa through eastern Africa.** Nat Genet 1999, **23**(4):437-441. [PubMed Abstract] [Publisher Full Text]

Rajkumar R, Banerjee J, Gunturi HB, Trivedi R, Kashyap VK: (2005) Phylogeny and antiquity of M macrohaplogroup inferred from complete mtDNA sequence of Indian specific lineages. *BMC Evol Biol*, **5**:26.

Rando JC, Pinto F, Gonzalez, AM, Hernandez M, Laruga JM, Cabrera VM, Bandelt H J . (1998) Mitochondrial DNA analysis of northwest African populations reveals genetic exchanges with European, near-eastern and sub-Saharan populations. **Ann Hum Genet**, 62: 531-550.

Roychoudhury S, Roy S, Basu A, Banerjee R,

Vishwanathan H, Usha Rani MV, Sil SK, Mitra M,

Majumder PP. (2001) Genomic structures and

population histories of linguistically distinct tribal

groups of India. **Hum Genet**, 109:339–350 First citation

in article | PubMed | CrossRef


Sastri, Nulakanta. (1966). **History of South India**.

Madras: Oxford University Press.


Sergent, Bernard (1992). *Genèse de L'Inde*. Paris:
Payot.

Stevanovitch A, Gilles A, Bouzaid E, Kefi R, Paris F,

Gaynaud RP, Spadoni I L. (2003) Mitochondrial DNA

Sequence Diversity in Sedentary Population from Egypt,

**Ann Hum Genet**,68:21-29.


Tambets K, Kivisild T, Metspalu E, Parik J, Kaldma K,

Laos S, Tolk HV, Gölge M, Demirtas H, Geberhiwot T,

Papiha SS, de Stefano GF, Villems R: **The topology of the**

**maternal lineages of the Anatolian and Trans-Caucasus populations and the peopling of Europe: Some preliminary considerations**. In *Archaeogenetics: DNA and the population prehistory of Europe*. Edited by Renfrew C, Boyle K. Cambridge, UK: McDonald Institute for Archaeological Research, University of Cambridge, 2000 :219-235

Sun C, Kong QP, Palanichamy MG, Agrawal S, Bandelt HJ, Yao YG, Khan F, Zhu CL, Chaudhuri TK, Zhang YP. (2006) The dazzling array of basal branches in the mtDNA macrohaplogroup M from India as inferred from complete genomes. **Mol Biol Evol**, 23:683-690.

Tanaka M, Cabrera VM, González AM, Larruga JM, Takeyasu T, Fuku N, Guo LJ, Hirose R, Fujita Y, Kurata M, Shinoda K, Umetsu K, Yamada Y, Oshida Y, Sato Y, Hattori N, Mizuno Y, Arai Y, Hirose N, Ohta S, Ogawa O, Tanaka Y, Kawamori R, Shamoto-Nagai M, Maruyama

W, Shimokata H, Suzuki R, Shimodaira H. (2004).
Mitochondrial genome variation in Eastern Asia and the
peopling of Japan. *Genome Res*, **14**:1832-1850.

Upadhyaya,P & Upadhyaya,S.P. (1979) .Les liens
entre Kerala et l"Afrique tels qu'ils resosortent des
survivances culturelles et linguistiques, **Bulletin de
L'IFAN**, no.1:.100-132.

Upadhyaya,P & Upadhyaya,S.P. (1976). Affinites
ethno-linguistiques entre Dravidiens et les Negro-
Africain, **Bull.de L'IFAN**, No.1: 127-157.

Verneaux,R: **Les Origines de l'humanite.** Paris: F.
Riedder & Cie, 1926.

Wai-Ogusu,A.(1973). Was there a Sangoan industry
in West Africa, West African Jour of Arcaheo, 3:191-96.

Welmers, W. (1971). Niger-Congo Mande, **Current**

**trends in Linguistics** 7 :113-140.

Wendorf,F. :**The History of Nubia**, Dallas, 1968.

Winters, C.(2007). Did the Dravidian Speakers Originate in Africa? **BioEssays**,27(5):497-498.

_____.(2008a). African millets carried to India by Dravidian Speakers ?

**Annals of Botany** (19 March 2008).

http://aob.oxfordjournals.org/cgi/eletters/100/5/903

_____(2008b) .Origin and Spread of Dravidian Speakers, Int. J. Hum Genet., 8(4):325-329.

Yao YG, Kong QP, Bandelt HJ, Kivisild T, Zhang YP: **Phylogeographic differentiation of mitochondrial DNA in Han chinese**. *Am J Hum Genet* 2002 , **70**:635-651.

# Chapter Four: The Kushite Spread of Haplogroup R1*-M173 from Africa to Eurasia

## Abstract

In this paper we discuss the role of the Kushites in the spread of R1*-M173. Human y-chromosome haplogroup R1*-M173 is mainly found in Africa. Haplogroup R1*-M173 is the pristine form of haplogroup R. In Africa researchers have detected frequencies as high as 95% among Sub-Saharan Africans. The phylogenetic, craniometric, textual, historical and linguistic evidence support the demic diffusion of Niger-Congo (Nilo-Saharan) carriers of R1*-M173 from Africa to Eurasia between 4-5kya.

Introduction

Archaeogenetics is the use of genetics, archaeology and linguistics to explain and discuss the origin and spread of homo sapien sapiens Renfrew, 2010). In this paper we will use archaeogenetics to examine and discuss the spread of haplogroup R-M173 by the ancient Kushites.

Researchers have outlined two possible out of Africa events in the past 40ky. Although these out of Africa events occurred during prehistory the Classical writers of Greece and Rome discussed a recent migration of people from Africa into Eurasia. This African population was called: Kushites.

Materials and Methods

A review of the archaeological, linguistic, genomic and craniometric literature was used to explore the role of the Kushites in the spread of haplogroup R from Africa to Eurasia. In this analysis of the linguistic, craniometric, and related scientific literatures we will determine if archaeological and genomic evidence can trace a migration event and dispersal of Kushites into Eurasia as maintained by the Classical writers. This study was conducted in Chicago at the Uthman dan Fodio Institute in 2009.

Results and Discussion

We analyzed the craniometric , linguistic, archaeological and y-chromosome sequences of African and Eurasian populations from the literature relating to these diverse fields.

This literature provides us with a critical

143

examination of the distribution of R1*-M173 . It

presents a genetic pattern of this haplogroup from

Africa to Eurasia, and the dispersal of a significant

African male contribution to Eurasia in the past

4ky.

The pristine form of R1*M173 is found only in

Africa (Cruciani et al, 2002, 2010). Haplogroup R1*-

M173 (xSRY 10831, M18, M117, M173, M269).

Haplogroup R-M173 is ancestral to R-P25 (xM269)

and other Eurasian downstream markers.

The Eurasian R1b y-chromosome has the M269

mutation. The R-P25* haplogroup has been found

in Europe, West and East Asia (Cruciani et al, 2010).

Figure 1 we see the frequency of R1*-M173 in

Africa and Eurasia. InThe frequency of Y-

chromosome R1*-M173 in Africa range between

7-95% and averages 39.5% (Coia et al,2005). The

R*-M173 (haplotype 117) chromosome is found

frequently in Africa, but rare to extremely low

frequencies in Eurasia. The Eurasian R haplogroup

is characterized by R1b3-M269. The M269 derived

allele has a M207/M173 background.

In Figure 1 we provide the frequencies of y-

chromosome M-173 in Africa and Eurasia. Whereas

only between 8% and 10% of M-173 is carried by

Eurasians, 82% of the carriers of this y-

chromosome are found in Africa.

Coia et al (2005) provides substantial data that

the presence of R1*-M173 did not follow the

spread of the spread of mtDNA haplogroup U6

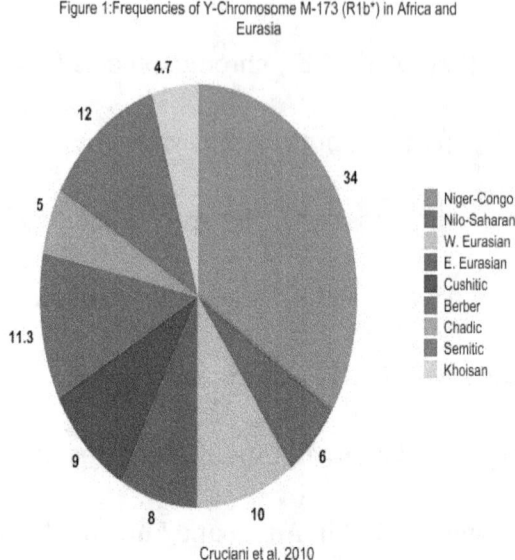

Figure 1:Frequencies of Y-Chromosome M-173 (R1b*) in Africa and Eurasia

Cruciani et al, 2010

in Sub-Saharan Africa, which is found in North

Africa (Coia et al, 2005). This suggest that R1*-

M173 may not be the result of back migration from

Asia if this theory depends on the spread of

haplogroup U6 in areas where R1*-M173 is found.

The majority of West Africans formerly lived just below Egypt in Nubia, before they moved westward into Cameroon, the Niger Valley and Senegambian regions. This part of Africa was inhabited by the Kushite people in ancient times.

The Kushite people are usually associated with the C-Group civilization of Nubia and Egypt. The center of their civilization was situated first in Wawat (southern Egypt) and later Kerma. The majority of West Africans speak languages that belong to the Niger-Congo group of languages. The Niger-Congo languages originated in Nubia and were probably spoken by some of the Kushites.

Wm. E. Welmers (1971),explained that the Niger-Congo homeland was in the vicinity of the upper Nile valley. He believes that the Westward migration from Nubia began 5000 years ago. This was the center of the C-Group civilization.

In support of this theory Welmers (1971) discusses the dogs of the Niger-Congo speakers. This is the unique barkless Basenji dogs which live in the Sudan and Uganda today, but were formerly recorded on Egyptian monuments (Welmers,1971). The Basanji dog is the Egyptian hieroglyphic sign for dog.

According to Welmers (1971) the Basanji, is related to the Liberian Basenji breed of the Kpelle and Loma people of Liberia. Welmers (1971) believes that the Mande took these dogs with them on their migration westward. The Kpelle and

Loma speak Mande languages.

Welmers (1971) believes that the Niger Valley region and other regions of West Africa may have been unoccupied when the Mande migrated westward Nubia. In support of this theory Welmers' notes that the Liberian Banji dogs ,show no cross-breeding with dogs kept by other African groups in West Africa, and point to the early introduction of this cannine population after the separation of the Mande from the other Niger-Congo speakers in the original upper Nile homeland for this population. As a result, he claims that the Mande migration occured before these groups entered the region.

Linguistic research make it clear that there is a close relationship between the Niger-Congo Superlanguage family and the Nilo-Saharan languages spoken in the Sudan. Heine and Nurse (2000), discuss the Nilo-Saharan connection. They note that when Westerman described African languages he used lexical evidence to include the Nilo-Saharan and Niger-Congo languages into a Superfamily he called "Sudanic" (Heine & Nurse, 2000). Using Morphological and lexical similarities Gregerson indicated that these languages belonged to a macrophylum he named " Kongo-Saharan" (Heine and Nurse, 2000). Research by Blench reached the same conclusion, and he named this Superfamily: "Niger-Saharan" (Heine & Nurse, 2000).The close relationship between Niger-

Congo and Nilo-Saharan suggest an intimate relationship formerly existed between the diverse speakers of these language families, probably in Nubia.

Genetic evidence supports the upper Nile origin for the Niger-Congo speakers. Rosa et al (2007), noted that while most Mande & Balanta carry the E3a-M2 gene, there are a number of Felupe-Djola, Papel, Fulbe and Mande carry the M3b*-M35 gene the same as many people in the Sudan.

In addition to haplogroup E3, we also find some carriers of haplogroup R1*-M173 in Egypt and the Sudan. In Figure 2 we observe that the majority of the carriers of y-chromosome M173 in Africa speak Niger-Congo languages. This genetic evidence makes it clear that R1*-M173 was probably carried

by some C-Group speakers before they migrated

out of the Upper Nile Valley region.

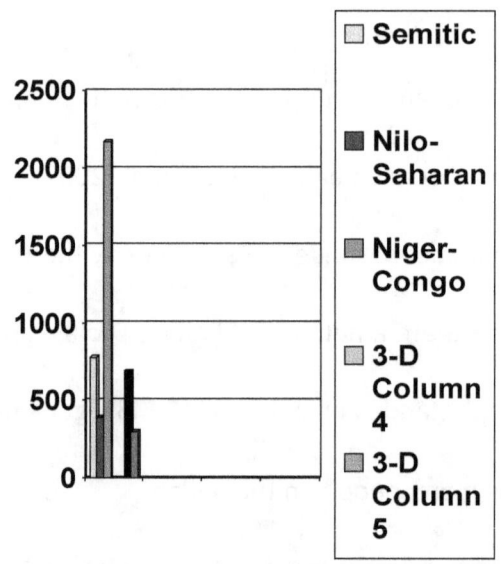

Figure 2:Frequencies of Y-Chromosome M-173 (R1b*) in Africa

Frequencies from Cruciani et al, 2010

Welmers (1971) proposed an Upper Nile

homeland for the Niger-Congo speakers. He claims

that they remained intact until 5000 years ago. This

view is supported by linguistic and genetics

evidence.

The Greco-Roman writers made it clear that

there were two Kushite empires one in Asia and

the other group in the area we call the Sudan

(Hansberry,1981). The Greek writer Homer alluded

to the two Kushite empires, when he wrote "a race

divided, whom the sloping rays; the rising and the

setting sun surveys". The Greek traveler/historian

Herodutus claimed that he derived this information

from the Egyptians.

The Kushites were also called Ethiopians. The

term Ethiopian comes from two Greek terms:

Ethios 'burnt' and ops 'face', as a result Ethiopian

means the 'burnt faces' (Winters,2005). Herodutus
and Homer, described these Ethiopians as "the
most just of men ;the favorites of the gods"
(Hansberry,1981). The classical literature makes it
clear that the region from Egypt to India was called
by the name Ethiopia.

Hansberry (1981) provides a great discussion
of the evidence of African Kushites ruling in Asia
and Africa. Some ancient scholars noted that the
first rulers of Elam were of Kushite origin.
According to Strabo, the first Elamite colony at
Susa was founded by Tithnus, a King of Kush.
Strabo in Book 15, Chapter 3,728 wrote that in fact
it is claimed that Susa was founded by Tithonus
Memnon's father, and his citadel bore the name
Memnonium. The Susians are also called Cissians.

Aeschylus, calls Memnon's mother Cissia.

The Elamite language is closely related to Dravidian (McAlpin, 1974, 1981 ;Winters, 1989) and Niger-Congo languages (Winters,1985a, 2005).

There is genetic, linguistic and archaeological evidence pointing to the African origin of the Dravidian speakers in India (Aravanan 1980; Winters 2007). Lal's (1963) research suggests that the Dravidian speaking people may have belonged to the C-Group. The C-Group people spread culture from Nubia into Arabia, Iran and India as evidenced by the presence of black-and-red ware (BRW). Although the Egyptians preferred the cultivation of wheat, many ancient C-Group people were agro-pastoral people who cultivated Millet/Sorghum and raised cattle. It was the

Dravidians who probably took millet to India (Winters 2008b).

The C-Group people used a common black and red ware that has been found from the Sudan, across Southwest Asia and the Indian Subcontinent all the way to China (Singh 1982). The earliest use of this BRW was during the Amratian period (c.4000 3500 BC). The users of the BRW were usually called Kushites.In Figure 4, we see the Kushite expansion from Africa to Asia.

Controversy surrounds the origin of the Dravidian languages. There is abundant evidence that the Dravidian languages are genetically related to the Niger-Congo group (Aravanan 1979, 1980 ; Upadhyaya and Upadhyaya, 1976, 1979; Winters 1985a, 1988, 1989).

The Proto-Dravidian speakers probably

migrated across Arabia to reach India. The first

civilization in Arabia was the Tihama culture. The

Tihama civilization probably originated in Nubia.

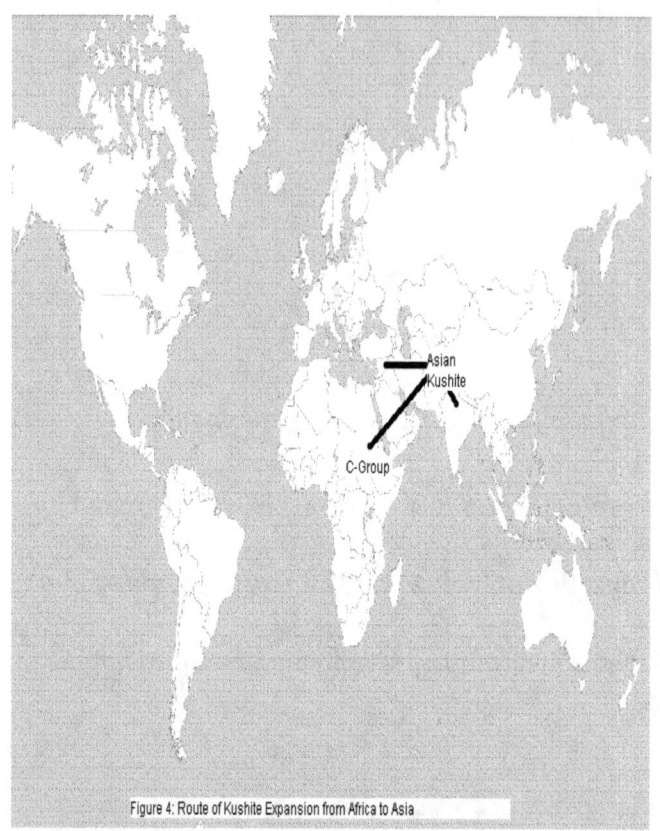

Figure 4: Route of Kushite Expansion from Africa to Asia

The Armenians made it clear that the ancients

called Persia, Media,Elam , Aria, and the entire area

between the Tigris and Indus river Kush

.Bardesones, writing in his Book of the Laws of

Countries, in the 2nd Century said that the

"Bactrians who we called Qushani (or Kushans)" (Winters,2000, 2005).The Armenians, called the earlier Parthian: Kushan and acknowledged their connection with them. Homer, Herodotus, and the Roman scholar Strabo called southern Persia AETHIOPIA (Hansberry, 1981). The Greeks and Romans called the country east of Kerma: Kusan.

The Kushites are associated with the C-Group people of Nubia, and the Kerma civilization. The Kushites practiced an agro-pastoral economy and they made a characteristic red-and-black pottery that they spread from Nubia to China.

Archaeologists agree that Black and red ware (BRW) indus unearth on many South India sites are related to Dravidian speaking people. The BRW style has been found on the lower levels of

Madurai and Tirukkampuliyur. B.B. Lal (1963) made

it clear that the South Indian BRW was related to

Nubian ware dating to the Kerma dynasty. Singh

(1982) made it clear that he believes that the BRW

radiated from Nubia through Mesopotamia and

Iran.

The legacy of the Kushites in Asia is evident in

the use of their ethonym as a place-name

characterized by the name Kush.

The Kushites when they migrated from Middle

Africa to Asia continued to call themselves

Kushites. This is most evident in place names and

the names of gods. The Kassites, chief rulers of Iran

occupied the central part of the Zagros

(Winters,2005). The Kassite god was called

**Kashshu**, which was also the name of the people (Winters, 2000). The **K-S-H**, name element is also found in India. For example **Kishkinthai**, was the name applied to an ancient Dravidian kingdom in South India. Lets not forget that the Kings of Sumer, were often referred to as the " Kings of Kush".

The major Kushite tribe in Central Asia was called Kushana. The Kushan of China were **Ta Yueh-ti** or "the Great Lunar Race". Along the Salt Swamp, there was a state called **Ku-Shih** of Tibet. The city of **K-san**, was situated in the direction of Kushan, which was located in the Western part of the Gansu Province of China (Winters, 2005).

Anatolia was occupied by many Kushite groups,

161

including the Kashkas and Hatti. The Hatti ,like the Dravidian speaking people were probably related . The Hatti were probably members of the Tehenu tribes.

The Tehenu were composed of various ethnic groups. The Tehenu was a major African population associated with the C-Group. One of the Tehenu tribes was identified by the Egyptians as the **Hatiu** or **Haltiu (** El Mosallamy,1986) .The Hatiu, may represent the Hatti tribe.

Singer (1981) has suggested that the Kaska, are remnants of the indigenous Hattian population which was forced northward by the Hittites. But at least as late as 1800 BC, Anatolia was basically settled by Hattians (Steiner , 1981).

We can use craniometric data to understand

ancient population history. The craniometric evidence indicates a process of demic diffusion of Kushite people into Mesopotamia and Anatolia between 5-4kya. Craniometric data sets support a continuos dispersal modal of Sub-Saharan Africans from Africa to Eurasia (Ricault & Waelkens,2008 ; Tomczyk et al 2010) between 5-4kya.

There is a positive relationship between crania from Africa and Eurasia. The archaeologist Marcel-Auguste Dieulafoy (Dieulafoy,2004) and Hanberry (1981) maintains that their was a Sub-Saharan strain in Persia . These researchers maintain that it was  evident that an Ethiopian dynasty ruled Elam from a perusal of its statuary of the royal family and members of the army ( Dieulafoy, 2004; Dieulafoy, 2010;Hansberry,1981).

Dieulafoy (2010 ) noted that the textual evidence and iconography make it clear that the Elamites were Africans, and part of the Kushite confederation .Dieulafoy (2010) made it clear that the Elamites at Susa were Sub-Saharan Africans.

Marcel Dieulafoy and M. de Quatrefages observed that the craniometrics of the ancient Elamites of Susa indicate that they were Sub-Saharan Africans or Negroes (Dieulafoy,2010).

Ancient Sub-Saharan African skeletons have also been found in Mesopotamia (Tomczyk et al, 2010). The craniometric data indicates that continuity existed between ancient and medieval Sub-Saharan Africans in Mesopotamia (Ricault & Waelkens,2008).

There is a genetic linguistic relationship

between the Dravidian, Elamite and Niger-Congo languages (McAlpin, 1974,1981; Winters,1989). The linguistic evidence makes it clear that a genetic relationship exist between Elamite and the Mande languages (Winters, 1985b,1989).

The relationship between the Mande and Elamite languages is interesting because the Garama or Garamante people of Crete, probably spoke a Mande language. Graves (1980) claimed that the Garamante formed part of the Mande group that live along the Niger River.

The relationship between the Elamite and Mande languages is interesting because Ricault and Waelkens (2008) noted a relationship between the Anatolia populations and Niger-Congo speakers. The Mande languages belong to the Niger-Congo

165

Superfamily of languages. This suggest that the Garamante spoke a Niger-Congo language.

The founders of civilization on Crete were the Garamante. The Minoans called themselves Keftiu. The Egyptians recorded some Keftiu names in their hieroglyphs. These names are common clan names among the Mande speaking people (Winters,2010).

Ricault and Waelkens (2008) provide craniometric and other evidence of a Cretan or Keftiu expansion into Anatolia. They believe that the Cretans colonized Anatolia; and that negro skeletons come from Illion-Troy, which as we discussed earlier was founded by Kushites (Winters, 2005). The research of Ricault and Waelkens (2008) is significant because they noted that the craniometric data set from Anatolia is related to West African (Niger-Congo) and Kerma (Kushite) populations.

Conclusion

The phylogenetic profile of R-M173 supports an ancient migration of Kushites from Africa to Eurasia as suggested by the Classical writers. In Figure 3, we outline the spread of haplogrorp R from Nubia into Asia and West Africa. This expansion of an African Kushite population probably took place Neolithic period.

The accumulated Classical literature, archaeological, craniometric, genetic and linguistic evidence suggest a genetic relationship between the Kushites of Africa and Kushites in Eurasia that can not be explained by microevolutionary mechanisms. The phylogeographic profile of R1*-

M173 supports this ancient migration of Kushites from Africa to Eurasia as suggested by the Classical writers. This expansion of Kushites into Eurasia probably took place over 4kya.

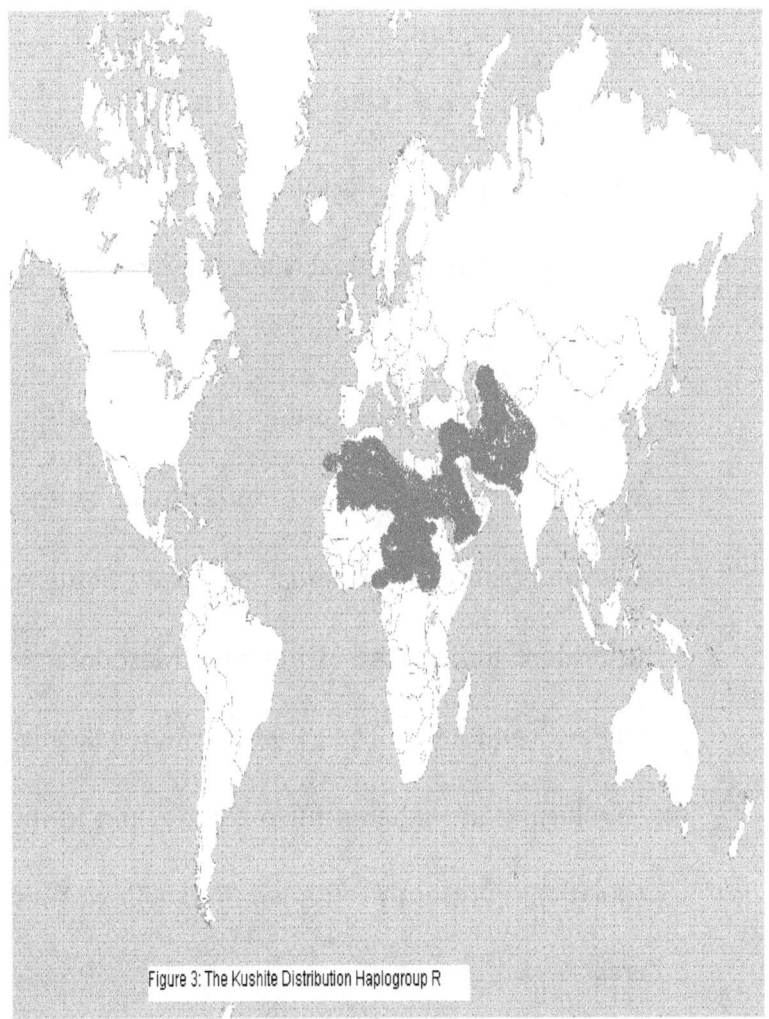

Figure 3: The Kushite Distribution Haplogroup R

The linguistic evidence makes it clear that the

Nilo-Saharan and Niger-Congo languages are

related. The genetic evidence indicates that Nilo-

Saharan and Niger-Congo speakers carry the y-chromosomes M3b*-M35 and R1*-M173, an indicator for the earlier presence of speakers of this languages in an original Nile Valley homeland.

The distribution of y-chromosome specific haplogroups in areas formerly occupied by the Kushite people of Asia reveal continuity between the ancient inhabitants of Anatolia, Mesopotamia and Persia and Africa. The genetic pattern indicates a significant Sub-Saharan male contribution to the populations presently situated in south-western Eurasia.

The tradition of a Kushite migration from Africa to Asia recorded in the classical literature is supported by the clinal biological pattern of y-chromosome lineages in Africa and Eurasia. The presence of R1*-M173 among Anatolians and Iranians supports a Neolithic demic diffusion of Kushite agropastoral populations into this region. The cranial discrete traits, y-chromosome haplogroups and linguistic affiliations shared between Sub-Saharan Africans, the ancient Mesopotamian, Anatolian and Iranian populations can only be the result of a human migration from Africa to Eurasia in ancient times as noted by the Classical writers of Greece and Rome.

References:

Aravanan KP .1979. Dravidians and Africans. Madras:Paari Nilayam.

Aravanan KP .1980. Notable Negroid elements in Dravidian India. J Tam Stud, 17: 20-45.

Coia V. , G Destro-Bisol, F Verginelli, C Battaggia, I Boschi, F Cruciani, G Spedini, D Comas and F

Calafell, 2005. Brief communication: mtDNA variation in North Cameroon: lack of Asian lineages and implications for back migration from Asia to sub-Saharan Africa, *Am J Phys Anthropol* (http://www3.interscience.wiley.com/cgi-bin/fulltext/110495269/PDFSTART) (electronically published May 13, 2005; accessed August 5, 2005).

Cruciani, F., Santolamazza,P., Shen, P., Macaulay, V., Moral P., Olckers,A. 2002. A Back Migration from Asia to Sub-Saharan Africa is supported by High-Resolution Analysis of Human Y-chromosome Haplotypes. *Am J. Hum Genet.,* 70:1197-1214.

Cruciani,F., Trombetta,B., Sellitto, D., Massaia,A. destroy-Bisol,G., Watson, E., Colomb, E.B. 2010. *Eur J. Hum Genet.,*doi:10.1038/ejhg.2009.231: 1-8.

Dieulafoy, J. 2004. *The Project Gutenberg EBook of*

*Perzi, Chaldea en Susiane*, by Jane Dieulafoy.

Retrieved 04/04/10

http://www.gutenberg.org/files/13901/13901-

h/13901-h.htm


Dieulafoy, M.A.2010.. *L'Acropole de Suse d'après*

*les fouilles exécutées en 1884, 1885, 1886, sous les*

*auspices du Musée du Louvre.* Retrieved 04/04/10

from :

http://www.archive.org/stream/lacropoledesused0

1dieu#page/2/mode/2up


Graves, R.1980. The Greek Myths. Middlesex:

Penguid Book. Ltd. 2vols.

Hansberry,L.1981. William Leo Hansberry: African

History Notebook.Vol 2. Africa & Africans as Seen

by Classical Writers, (Ed.) by Joseph E. Harris.

Washington,D.C.: Howard University Press. ISBN:0-

88258-0037.

Heine,B. and Nurse,D. (Eds.).2002. *African*

*languages: An introduction* , Cambridge University

Press. ISBN 0-521-666295.


Lal BB. 1963. "The Only Asian Expedition in

threatened Nubia: Work by an India Mission at

Afyeh and Tumas". *The Illustrated Times,* London

20 April.

McAlpin DW 1974. "Toward Proto Elamo

Dravidian". Lang. **50(1):** 89-101.

McAlpin DW 1981. Proto Elamo Dravidian: The

evidence and its implications. *Trans of the Am Philo*

*Soc,* 71, Part 3: Philadelphia.

El Mosallamy,A.H.S. (1986). Libyco-Berber relations with ancient Egypt:The Tehenu in Egyptian records. In (pp.51-68), p.55; and L. Borchardt, Das Grabdenkmal des Konigs Sahure. Vol. II, Table 1.

Rosa A, Ornelas C, Jobling MA, Brehm A, Villems R. Y-chromosome diversit6y in the population of Guinea-Bissau: a multiethnic perspective, BMC Evol Biology 2007; 7, 124.

http://www.ncbi.nlm.nih.gov/pmc/articles/PMC1976131/?tool=pubmed

Renfrew, C. 2010. Archaeogenetics—Towards a 'New Synthesis'? *Cur Biol,* (February 23, 2010) 20:R162-R165.

Ricaut,F.X. and Waelkens.2008. Cranial Discrete Traits in a Byzatine Population and Eastern

Mediterranean Population Movements, Hum Biol, 80(5):535-564.

Singh, H.N. 1982. *History and archaeology of Blackand Red ware.* Vedic Books.net: Manchester.

Singer, I. (1981). Hittites and Hattians in Anatolia at the beginning of the Second Millennium B.C., J of Indo-Euro Stud, 9 (1-2):119-149.

Steiner, G. (1981). The role of the Hittites in ancient Anatolia, J of Indo-Euro Stud, 9 (1-2): 150-169.

Tomczyk,J., Jedrychowska-Danska, K., Ploszaj,T & Witas H.W. (2010). Anthropological analysis of the osteological material from an ancient tomb (Early Bronze Age) from the middle Euphrates valley, Terqa (Syria) , International Journal of Osteoarchaeology, Retrieved 04/04/10 from (www.interscience.wiley.com)DOI:10.1002/oa.115 0.

Upadhyaya P, Upadhyaya SP 1979. Les liens entre Kerala et l''Afrique tels qu'ils resosortent des survivances culturelles et linguistiques. *Bull. de L'IFAN,* **1:** 100-132.

Upadhyaya P, Upadhyaya SP 1976. Affinites ethnolinguistiques entre Dravidiens et les Negro-Africain.*Bull.de L'IFAN,* **1:** 127-157.

Welmers, Wm E .1971. "Niger-Congo Mande", *Cur trends in Ling* 7:113-140.


Winters,Clyde Ahmad. 1985a. "The genetic Unity between the Dravidian ,Elamite, Manding and Sumerian Languages", P Sixth ISAS ,1984, (Hong Kong:Asian Research Service) pages 1413-1425.

Winters CA 1985b. The Proto Culture of the Dravidians, Manding and Sumerians. *Tam Civ,* **3(1)**:I 9.

Winters,C. A. 1989."Tamil,Sumerian and Manding and the Genetic Model",Int J of Dra Ling,18(I): 98-127..

Winters, C.A. 2000. "Memnonia ". In Shades of Memnon, Book II. Brother G (pp.13-33).SekerNefer Group,Chicago. ISBN 0-9662374-2-0

Winters,C.A.2005. Afrocentrism : Myth or Science. Lulu.com. ISBN 978-1-4116-5276-7

Winters CA 2007. Did the Dravidian Speakers Originate in Africa? *BioEssays,* **27(5):** 497-498.

Winters CA 2008a. Can parallel mutation and neutral genome selection explain Eastern African M1 consensus HVS-1 motifs in Indian M

Haplogroups. *Int J Hum Genet*, **13(3):** 93-96.

Winters CA 2008b. African millets taken to India by Dravidians. *Ann of Bot,* h **Winters,C.A. 2010. The Ancient Minoans: Keftiu were Mande Speakers .**

**Retrieved 04/12/2010 from**

http://bafsudralam.blogspot.com/search?q=Minoa nsttp://aob.oxfordjournals.org/cgi/eletters/100/5/903#49

# Chapter Five: Inference of Ancient Black Mexican Tribes and DNA

## Abstract

*Background*: Controversy surrounds the time period when Indigenous Mexican-African admixture occurred. Most researchers assume this admixture took place after the Atlantic Slave Trade. But, Spanish eyewitness accounts, Mayan skeletons with sickle cell anemia, and West African skeletal remains generally, indicate that there were Black Native Mexican and Meso-American communities in Meso-America before 1492. Using genetic association studies of available Indigenous Mexican and African genome-wide SNP genotypes and HLA we infer the probable pre-or post Columbian date for the admixture. Here we analyze the historical and archaeogentic literature relating to the American foundational haplogroups and HLA to extract ancestry information detailing when Indigenous Mexican-African admixture took place.

*Results:* Indigenous Mexican and African archaeogenetic, DNA and HLA resources were analyzed to determine to what extent admixture had occurred between these populations. The sample indicated that Indigenous Mexican-African admixture has taken place across Mexican fundamental male and female lineages; and that Africans and Indigenous Mexicans share HLA alleles. In addition, archaeogenetic evidence including, African [Mande] inscriptions, Mande substratum in Mayan languages, Africans depicted in Mayan murals at San Bartolo and Xultun, African skeletons generally, and ancient Mayan skeletons with sickle cell anemia support Spanish eyewitness accounts of Black Native American tribes [Otomi, Chontal (Mayan speaking group) ,Yarura and etc.] in Meso-America when they arrived on the scene.

*Conclusion*: We demonstrate that given the age of the African skeletons, excavated at Meso-American archaeological sites and Spanish eyewitness accounts of Black Mexicans, Indigenous Mexican- African admixture occurred prior to the European discovery of America. The date for the African skeletons indicate that there were several waves of West Africans who probably introduced African haplotypes into the Americas. The 25,000 Malians who sailed to America in 1310 probably had a major influence on the exchange of African genes in the Americas.

**Key Terms:** haplogroup, Mali, Black Native Americans, Meso-America, y-Chromosome, HLA, Indigenous Mexicans

## Background

Meso-America is the geographical name for Mexico and the countries of Central America. Today people believe that the Blacks of Mexico and the Blacks of Guatemala, Hondurus and Belize are the descendants of Sub-Saharan African (SSA) slaves taken to Mexico during the Atlantic Slave trade.

Researchers have suggested that Sub-Saharan Africans (SSA) were among the first Americans (1-6). Spanish explorers found Sub-Saharan African communities in Mexico when they arrived (1,7).

Sub-Saharan Africans were living in Mexico in 1492 (1-2). These SSA were trading with the mongoloid Amerindians, in addition to having their own settlements in the Americas. Amerigo Vespucci met African merchants on their way back to West Africa in the Middle of the Atlantic Ocean (7).

Much of what we know about African nautical sciences comes from Vaco da Gama. Vasco da Gama is said to have found information about the West Indies from Ahmad b. Majid, a West African he met during his travels along the West Coast of Africa (8). Da Gama claimed that ibn Majid wrote a handbook of navigation on the Indian Ocean, the Red Sea, the Persian Gulf, Sea of Southern China and the waters around the West Indian Islands. Majid is also said to be the inventor of the compass (8-9).

The Spanish left us mention of many Sub-Saharan Communities in Central America and Mexico (10-11). These dark skinned Indians were Africans not mongoloid Indians. Paul Gaffarel noted that when Balboa reached America he found "negre veritables" or true Blacks(12). Balboa noted "...Indian traditions of Mexico and Central America indicate that Negroes were among the first occupants of that territory" (12)." This is probably why so many Mexicans have "African faces ".

In addition, eyewitness accounts of SSA populations in the Caribbean, and Mexico anthropologists have found SSA skeletons at Pre-Columbian sites (5, 13-17 ). Some of the ancient Mayan skeletal remains indicate that they suffered from sickle cell anemia an illness associated with Sub-Saharan Africans (18-20). The presence of sickle cell anemia among the ancient Maya, supports Quatrefages claim that the Chontal Maya were Africans( 7,11) .

**Method**

The research design used in this study is a literature based research methodology. We analyzed the DNA literature relating to the subclades of Indigenous foundation haplogroups and HLA, and compared them to African haplogroups. The sample includes genomic and HLA literature of Indigenous Mexicans and West African test populations.

An inter-population comparison of Indigenous Mexican and African genomic and HLA literature was conducted to make a database of shared HLA alleles and population admixture frequencies. Data mining of the literature was used to determine haplogroup and HLA frequencies presented in this study.

## Results

There is a high frequency of African-Mestizo admixture ranging between 20-40% (21). The admixture rate between Africans and indigenous Mexican Indians ranges between 5-50% (22-23).

The Amerindian haplogroups (hg) are descendant from the L3(M,N, & X) macrohaplogroup): ABCDN and X. The L3 (M,N,X) marcogroup converge at np 16223.

The mtDNA haplogroups ABC and X are subclades of haplogroup N. In Table 1, we see the distribution of haplogroup N, in the Americas.

The phylogeography of haplogroup C suggest that this American founder haplogroup differentiated in Siberia—Asia (24). The situation is not so clear for haplogrop B2, but A2 and D1 probably differentiated after the mongoloid Native American lineages diverged after crossing the Beringa Straits (24)

Haplogroup A2 has the motif 16111T,16223c, 16290T, 16319A and 16223C (25). Haplogroup A is rare in Siberia (26). Interestingly, haplogroup A absent in western North America is common in parts of Central America and Northern America where the Spanish reported the existence of Black Native American communities(1-2).

In a recent study of post-Classic Mexicans at Tlatilco, dating between 10-13 centuries the subjects carried the founder haplogroups A (36%), B (13%), C (4.3%) and D (17.4%) (27). We should note, that in Yucatec, the Mayans were predominately haplogroup A, the Maya in Hondurus, a stronghold of the Black Native Americans belonged to haplogroup C.

The mtDNA haplogroup A common to Mexicans is also found among the Mande speaking people and some East Africans (28-29). Haplogroup A found among Mixe and Mixtecs (28).The Mande speakers carry mtDNA haplogroup A, which is common among Mexicans (30). In addition to the Mande speaking people of West Africa, Southeast Africa Africans also carry mtDNA haplogroup A (29).

The major American Indian male lineages include R1, C,D and Q3.There is evidence of African admixture in the American y-chromosome haplogroups. The Q y-haplogroup has the highest frequency among indigenous Mexicans. The frequency hg Q varies from a high of 54% for Q-M243, and a low of 46% for QM (34).

Underhill et al , noted that:" One Mayan male, previously [has been] shown to have an African Y chromosome" (31). This is very interesting because the Maya language illustrates a Mande substratum, in addition to African genetic markers (3).

African y-chromosome are associated with YAP+ and 9bp. The YAP-→ associated with A-→G transition at DYS271 is found among Native Americans. The YAP+ individuals include Mixe speakers (32-33). YAP+ is often present in haplogroups (hg) C and D.

The DYS271 transition is of African origin (32).The DSY271 Alu insertion is found only in chromosomes bearing Alu insertion (YAP+) at locus DYS287 (33). The DYS271 transition was found among the Wayuu, Zenu and Inzano. The Mexican Native American y-chromosome bearing the African markers is resident in haplogroups C and D (34).

African ancestry has been found among indigenous groups that have had no historical contact with African slaves and thus support an African presence in America, already indicated by African skeletons among the Olmec and Mayan people. Lisker et al, noted that "The variation of Indian ancestry among the studied Indians shows in general a higher proportion in the more isolated groups, except for the Cora, who are as isolated as the Huichol and have not only a lower frequency but also a certain degree of black admixture. The black admixture is difficult to explain because the Cora reside in a mountainous region away from the west coast" (22).

A recent study of African – Mexican admixture yielded a frequency range between 22-41% (25). In one study the researcher found that 3% of Native Americans

showed African haplogroups (25). The African
haplogroups among indigenous Mexicans include
L0a1a'3, L2a1, L3b, L3d, and U6a (25). Interestingly, an
individual at Laguna de los Condores, Peru dating
between AD 1000-1500 carried L3 (36). Green et al also
found Indians with African genes in North Central
Mexico, including the L1 and L2 clusters (25).

An important indicator of African admixture is 9bp
(22,27). Haploroup B is defined by 9bp (27). Liinked to
haplogroup A.The The 9bp marker is also 9 base pair
marker is reported among the North Mexicans, and is
common among the Mixtec (27).

Some indigenous Mexicans show the G6PD deficiency.
In a study of Yucatecos, Tzellzal-Tzoltzil, Mixteca and
Mestizo it was found that people on the Oaxaca coast
suffered from G6PD deficiency (22). Lisker also found
G6PD deficiency in Costa Chica (22). The G6PD
deficiency is usually carried by SSA.

Indigenous Indians at Tlaxcala contains 8% African
genes, but historically no Africans lived in the area (37).
Researchers have also found L1, L2 & L3 clusters
among many Mexicans including the Cora, Mixtec and
Zapotecs (25,36, 39-41)

It is interesting to note that the proportion of African
haplotypes was roughly equivalent to the proportion of
European haplotypes [among North Central Mexican
Indians] cannot be explained by recent admixture of

African Americans for the United States (41). This is especially the case for the Ojinaga area, which presently is, and historically has been largely isolated from U.S. African Americans.

Table 1: <u>Native American Carriers of Haplogroup N</u>

| Geographical Area | Frequency |
|---|---|
| Na-Dene- North | 5.3% |
| Na-Dene South | 2.7% |
| Northern Amerindians | 8.0 |
| Central American | 0.4 |
| Southern American | 1.2% |
| Fuego-Patagonia | 1.7% |

In the Ojinaga sample set, the frequency of African haplotypes was higher than that of European hyplotypes"(41).

Human Leukocyte Antigens (HLA) polymorphism is used to investigate ethnic relationships and origins. Africans and Indigenous Mexicans share HLA alleles. In Table 2 we outline the relationship. Gutherie in a study of the HLAs in indigenous American populations, found that the V antigen of the Rhesus system, considered to be an indication of African ancestry, is found among Indians in Belize and Mexico centers of Mayan civilization and where the Spanish reported the existence of Black Native American communities (45). Dr. Gutherie also noted that A*28 common among Africans has high frequencies among Eastern Maya (45).

In addition to A*28 , there is a high frequency of HLA B*35 among Mexicans and SSA (46). The frequency of HLA B*35 among indigenous Mexicans and SSA is high ranging between 22-31% among SSA populations and 30-45% among MA groups (46). It is interesting to note that the Otomi, a Mexican group identified as being of African origin and six Mayan groups show the B Allele of the ABO system that is considered to be of African origin.

R-M173 is also found in Mexico. Haplogroups R and Q are part of the CT macrogroup which dates back 56kya. Haplogroup R branches from hg Q, with the SNP M242.

The CT haplogroup has SNP mutation M168, along with P and M294. Haplogroup P (M45) has two branches Q (M242) and R-M207 which share the common marker M45.

The M45 chromosome is subdivided by the biallelic variant M173 (35). In Africa we find P (M173), R1b (M343) and V88; and R1b1a2 (M269).

Native Americans carry a high frequency of R-M173 (1).

The predominate y-chromosome in North America is R-M173 . R-M173 is found only in the Northeastern United States along with mtDNA haplogroup X (25%). Both haplogroups are found in Africa, but are absent in Siberia.

There are varying frequencies of y-chromosome M-173 in Africa and Eurasia. Whereas only between 8% and 10% of M-173 is carried by Eurasians, 82% of the carriers of this y-chromosome are found in Africa.

The pristine form of R1*M173 is found only in Africa (47-48). The frequency of Y-chromosome R1*-M173 in Africa range between 7-95% and averages 39.5% ( 49). The R*-M173 (haplotype 117) chromosome is found frequently in Africa, but rare to extremely low frequencies in Eurasia. The Eurasian R haplogroup is characterized by R1b3-M269. The M269 derived allele has a M207/M173 background.

Henn et al was surprised by this revelation of R-M269 among this Khoisan population (50). As a result, he interviewed the carries of R1b1b2a1a, and learned that no members of their families had relations with Europeans. The presence of R lineages among HG populations is not new. Wood et al reported Khoisan carriers of R-M269 (51 ). Bernielle-Lee et al , in their study of the Baka and Bakola pygmies found the the R1b1* haplogroup ( 52). These researchers made it clear that the Baka samples clustered closely to Khoisan samples (52).

The most common R haplogroup in Africa is V88. Given
the interaction between hunter-gatherer (HG) groups
and agro-pastoral groups they live in close proximity
too, we would assume that African HG would carry the
V88 lineage. Yet, as pointed out above the HG
populations carry R-M269 instead of V88 ( 53). The
implications of R-M269 among HG populations, and
Henn et al's shared African HG genome suggest that R-
M269 may represent a HG genome  thus an ancient
African R lineage. The presence of R-M269 among HG
human groupings fails to support a back migration of R-
M269 from Europe.

In a recent article on the R1 clade, Gonzalez et al , argue
that R1 probably spread across Europe from Iberia to
the east given the distribution of R1 in Africa (54).
Gonzalez et al   found that 10 out of 19 subjects in the
study carried R1b1-P25 or M269 as opposed to V88 in
Equatorial Guinea (54). This is highly significant because
it indicates that 53% of the R1 carriers were M269. This
finding is further proof of the widespread nature of this
so-called Eurasian genes in Africa among populations
that have not mated with Europeans.

This is very interesting given the presence on R-M173 is
found among many American Indian groups . R-M173
among  the North American Algonquian group range
from Ojibwe (79%), Chipewyan (62%), Seminole (50%),
Cherokee (47%), Dogrib (40%) and Papago (38%) (53) .
These Indian groups have a long association with
Africans and many live in areas were Europeans found

Black Native Mexicans.

The R haplogroup is carried by Mexicans. The frequency of hg R varies from Tarahumara (5.6%), Otomi (14.3%), Yucateca Maya (10.5%). There is also a high frequency of haplogroup R among the Ch'ol and Chontal which stood around 15% (38). The Ch'ol and Chontal also carry E1b1b (38).

## Discussion

Africans and indigenous Mexican admixture ranges between 5-50% (22-23). This is interesting because Indigenous Mexicans who have not lived in recent contact to Africans also have high admixture rates.

Some of the American foundational haplogroups have characteristic African YAP+ and 9bp. The YAP+ associated with A⬚G transition at DYS271, and is found among Native Mexicans.

Indigenous Mexicans carry R-M173. It is found in high frequency among the Ch'ol and Chontal which stood around 15% (38). The Ch'ol and Chontal also carry E1b1b (38).

HLA data indicate a high frequency of Indigenous Mexican-African admixture exist between these populations as seen in Table 2. The usual answer to this phenomena was that the admixture occurred as a result of the Atlantic Slave Trade, which led to millions of Sub-Saharan Africans being sold in Mexico.

Table 2: <u>Shared African and Indian HLA-A-B Alleles</u>

| HLA Allele | Amerindian* | Gambian ** | Malian ** | Nigerian ** |
|---|---|---|---|---|
| A*02 | 5% | 14.1% | 15.9 | |
| A*24 | 5% | 4% | 1.2% | 2.1% |
| A*31 | 5% | .03% | .018% | .036% |
| B*35 | 5% | 16.1% | .82% | .59% |
| B*40 | 5% | .06% | ____ | .09% |
| *Arnaiz-Villena et al, (43)  **Allsopp et al, (44) | | | | |

The problem with this solution to the African-Indigenous Mexican admixture is that the vast majority of African slaves worked on the East Coast of Mexico. Although this was the case we find that many Mexican Indigenous populations, living in other parts of Mexico show African admixture like the Cora, Huichol, Ojinaga and Tlaxcala Indians that did not live in association with

African slaves. This suggest that the Indigenous Mexican-African admixture may have taken place prior to the Africans slave trade.

Indigenous Mexicans probably mated with Black Native Americans and Sub-Saharan Africans. The majority of Black Native American tribes according to Quatrefages in The Human Species, include the Choco, Manabis, Yaruras, Guarani, Charruas, Othomi (Otomi), Yamassi, Tzendal/Chontal, the Mandinga (a member of the Cunan group of Mexico), the Blacks of Quareca and numerous tribes along the Orinoco river in Venezuela and the Isthmus of Darien (11). He also mentioned the Black tribes of the United States southwest including a tribe reported by Cabeza de Vaca called Mandicas (< Mandinka) (4).

The Otomi and Caribe spoke a Manding language (3) . The major center for the Manding was Panama. The major Amerindian group in this area was the Cunan group (3-4).

Some of the Indigenous Mexican-African admixture is probably the result of the Mali discovery of America (11). Mali was an ancient West African Empire which stretched from the Western Sudan to the Atlantic shore. This was a part of West Africa, that had a long tradition of nautical sciences (8-9).

Around A.D. 1310, thousands of Manding speakers arrived in the Americas from ancient Mali. Ibn Fadlullah al-Umari, in his encyclopedia *Masalik al Absar*, said the mariners from Mali during the reign of Abubakari made transatlantic voyages. Al-Umari, obtained his information from Mansa Musa, who was handed the kingship of Mali by Abubakari when he set out to colonize the Americas.

Ibn al-Umari wrote "But the Emperor [Abubakari] did not believe him", continued Musa, "He equipped two thousand vessels, a thousand for himself, and a thousand for water and supplies. He conferred power on me [Mansa Musa] and left with his companions on the ocean".

The expeditionary force of Mansa Abubakari, must have been immense, because the average boat on the Niger, in the 1500's A.D., could carry 80 men. This means that anywhere between 25,000 to 80,000 men may have sailed from Mali along with Mansa Abubakari. West African nautical science suggest that many members of Abubakari's expeditionary force made it to America, because we find many Mande (Malian) inscriptions throughout Meso-America, North America and South America (7).

It is obvious that even the founder American haplogroups show African admixture. Moreover, the evidence that the Mande and Mexicans carry mtDNA haplogroup A adds considerable weight to the idea that Africans mixed with indigenous Mexicans before 1492.

A pre-Columbian date for the admixture between Indigenous Mexicans and Africans is supported by archaeological, anthropological, eye witness accounts and linguistic evidence that Sub-Saharan Africans were already in America when Europeans first reached these shores.

The genetic evidence of Indigenous Mexican-African admixture is compelling. The frequency of HLA B*35 at 45% is highest among the Maya. We also find that the YAP+ associated with A→G transition at DYS271 and 9bp also has a high frequency among the Maya, all these markers are associated with African ancestry. This is not surprising because Quatrefages classified the Chontal Maya as Black Native Americans (3,7,11), and sickle cell anemia is found among ancient Mayan skeletons (18-20).

Archaeologists have found numerous SSA or Black Native American skeletons in Mexico (13-16), some of them showing evidence of sickle cell anemia (18-20). This suggest that Africans were in Mexico before the advent of Christophe Columbus and other Europeans(1-4,7). It also indicates that a large number of Blacks lived in Mexico before Columbus arrived in the Western hemisphere. This along with the Lancadon and Otomi being classified as Negroes (7,10-11), may indicate that contact between Indigenous Mexicans and Sub-Saharan African populations prior to the European discovery of America is the reason for the DNA admixture.

197

References

1.Alcina-Franch J.(1985). Los orígenes de America. :
Editorial Alhambra.

2. Arnaiz-Villena,A, Moscoso, J.,Serrano-Vela,I.
(2006).The uniqueness of Amerindians according to HLA
genes and the peopling of the Americas.
http://www.inmunologia.org/Upload/Articles/6/7/678.
pdf

3.Winters,C. ( 2011 ). Olmec (Mande) Loan Words in the
Mayan, Mixe-Zoque and Taino Languages. Current
Research Journal of Social Sciences 3(3): 152-179.

4. Winters,C. (2013). African Empires in Ancient
America. Createspace,Amazon.com.

5. Winters,C.(2015). Olmec Language and Literature.
Createspace,Amazon.com.

6. Winters,C. (2014). History of Blacks in America from
Pre-History to 1877. Createspace,Amazon.com.

7. Winters,C.(1977). Islam in Early North and South
America. Al-Ittihad, (July-October) pp.57-67.

8.Bazan, R.A.G. (1967). Latin America the Arabs and
Islam, Muslim World, pp.284-292.

9.Ferrand,G. (1928). Introduction a l'astrnomie nautique

des Arabes, Paris.

10. Orozco y Berra,M (1880). Historia Antigua y de la conquista de Mexico. https://archive.org/details/historiaantigua06berrgoog

11. Quatrefages, A de.(1889) . Introduction a L'Etudes des Races Humaines.

12. Gaffarel,P. (2010). Etude Sur Les Rapports De L'Amerique Et De L'Ancien Continent Avant Christophe Colomb.

13. Marquez,C.(1956). Estudios arqueologicas y ethnograficas. Mexico.

14. Wiercinski, A.(1969). Affinidades raciales de algunas poblaiones antiquas de Mexico, Anales de INAH, 7a epoca, tomo II, 123-143.

15. Wiercinski,A. (1972). Inter-and Intrapopulational Racial Differentiation of Tlatilco, Cerro de Las Mesas, Teothuacan, Monte Alban and Yucatan Maya, XXXIX Congreso International de Americanistas, Lima 1970 ,Vol.1, 231-252.

16. Wiercinski,A. (1972b). An anthropological study on the origin of "Olmecs", Swiatowit ,33, 143-174.

17. Wiercinski, A. & Jairazbhoy, R.A. (1975) "Comment", The New Diffusionist,5 (18),5.

18. Moore,S. (1929). The Bone Change in Sickle Cell

Anemia with A Note on Similar Changes Observed in Skulls of Ancient Mayan Indians, Journal of Missouri Medical Association, 26:561

19. Wailoo, Keith. (2002). Drawing Blood: Technology and Disease Identity in Twentieth-Century America. JHU Press.

20. Whittington, S. L., & Reed, D. M. (1997). Bones of the Maya: Studies of ancient skeletons. Washington, D.C: Smithonian Institution Press.

21. Lisker R, et al.(1996). Genetic structure of autochthonous populations of Meso-america:Mexico. Am. J. Hum Biol 68:395-404.

22. Suarez-Diaz,E. (2014) Indigenous populations in Mexico. Medical anthropology in the Work of Ruben Lisker in the 1960's. Studies in History and Philosop-hy of Biological and Biomedical Sciences 47:108-117.

23. Lisker,R.(1981. Estructura genetia de la poblacion Mexicana. Aspectos Medicos y Anthropologica, Mexico: Salvat.

24. Kumar S, Bells M Z, Melton P E, Blangero J, Currah J E. (2011). Large scale mitochondrial sequencing in Mexican Americans suggest a reappraisal of Native American Origins. BMC Evolutionary Biology 11:293.

25. Guardado-Estrada M, Juarez-Torres, E., Medina-Martinez I.(2009). A great diversity of American

mitochondrial DNA ancestry is present in the Mestizo population. Jour of Hum Genetics, 54:695-705.

26. Laluezza C, Perez-Perez A, Prats E1997. Lack of Found American Mitochondrial DNA Lineages in extinct aborigines from Tierra del Fuego-Patagonia. Hum Molecular Genet, 6(1)41-46.

27. Kemp B M, Resendez A, Roman J A, Berrelleza R, Malhi R.S.

28. Bonilla C, Gutierrez G, Parra E J, Kline C, Shriver M D. (2005). Admixture of a rural population of the State of Guerrero,Mexico, Am J Phys Anthropol. Dec;128(4):861-9.

29. Salas A, Richards M, De la Fe T, Lareu M V, Sobrino B, Sanchez-Diz P, Macaulay V, Carracedo A. (2002). The making of the West African mtDNA Landscape, Am J. Hum. Genet, 71:1082-1111.

30. Jackson B A, Wilson J L, Kirbah S, Sidney S S, Bassie L, Alle J A D, McLean D C Garvey W T.(2005). Am J Phys Anthropol. 128:156-163.

31. Underhill,P.A.,Jin,L., Zemans,R., Oefner,J and Cavalli-Sforza,L.L.(1996, January). A pre-Columbian Y chromosome-specific transition and its implications for human evolutionary history, Proceedings of the National Academy of Science USA,93, 196-200.

32. Lell, J T. (1997) Y chromosome polymorphisms in

Native Americans and Siberian populations: Identification of Native American y chromosome haplotypes. Hum Genet, 100(5-6):536-543.

33. Ruiz-Linares, A. (1999).Microsatellite provides evidence for y-chromosome diversity among the founders of the New World. Proc Natl Acad. Sci USA. 96(11):6312-6317

34. Branshi N O. (1997). Origin of Amerindian y-chromosome as inferred by the analysis of six polymorphism markers. Am J. Phys, Anth, 102(1)79-89.

35. Lell, T. (2002). The Dual Origin and Siberian affinities of Native American y-Chromosome. Am J Hum Genet., 70(1)196-206.

36. Genetic Geneology Tools: Ancient DNA. Retrieved 3/12/2015 from : www.y-str.org/p/ancientdna.html

37.Crawford,M.(2001). The Origins of Native Americans: Evidence from Anthropological Genetics. Cambridge University Press.

38. Quito A, Meraz M A, Camacho R, Schurr T, Vilar M(2013). Y-Chromosome diversity in Mayan Ch'ol and Chontal populations from Campeche and Tabasco. Retrieved 1/21/2015 from: http://www.ashg.org/2013meeting/abstracts/fulltext/fl 130123072.htm

39. Angelica Gonzalez-Oliver et al. (2001). Founding

Amerindian mitochondrial DNA lineages in ancient Maya from Xcaret, Quintana Roo. Am. Jour of Physical Anthropology, 116(3):230-235. Retreived 2/9/2006 at: http://www3.interscience.wiley.com/cgi-bin/abstract/85515362/ABSTRACT?CRETRY=1&

40. Maere Reidla, et al. (2003).Origin and Diffusion of mtDNA Haplogroup X Am J Hum Genet. 2003, 73(5): 1178–1190.

41. Green, L.D. (2000) mtDNA Affinities of the Peoples of North-Central Mexico. The American Journal of Human Genetics, 66(3) 989-998

42. Winters,C. (2011a). Comment: Genetic Evidence of Early Migrations into America. Retrived 2/18/2015: http://www.plosone.org/annotation/listThread.action?root=18395

43. Arnaiz-Villena A, Vargas-Alarcón G, Areces C, Enríquez-de-Salamanca M, Abd-El-Fatah-Khalil S, Fernández-Honrado M, Marco J, Martín-Villa JM, Rey D.(2014). Mixtec Mexican Amerindians: an HLA alleles study for America peopling, pharmacogenomics and transplantation. Immunological Investigations 43(8):738-55.

44. Allsopp, C.E.,R M Harding, C Taylor, M Bunce, D Kwiatkowski, N Anstey, D Brewster, A J McMichael, B M Greenwood, A V Hill.(1992). Interethnic genetic differentiation in Africa: HLA class I antigens in The Gambia. American Journal of Human Genetics, 50(2):

411–421.

Guthrie,J.L. (2006). Human lymphocyte antigens:Apparent Afro-Asiatic, southern Asian and European HLAs in indigenous American populations. Retrieved 3/3/2006 at:
http://www.neara.org/Guthrie/lymphocyteantigens02.htm

46. Winters,C. (2014) HLA-B*35 in Mexican Amerindians and African Populations. Forthcoming: Indian J Fundamental and Applied Life Scieces.

47. Cruciani,F., Trombetta,B., Sellitto, D., Massaia,A. destroy-Bisol,G., Watson, E., Colomb, E.B. (2010) European Journal of Human Genetics,(6 January 2010) doi:10.1038/ejhg.2009.231: 1-8.

48, Cruciani, F., Santolamazza,P., Shen, P., Macaulay, V., Moral P., Olckers,A. (2002) A Back Migration from Asia to Sub-Saharan Africa is supported by High-Resolution Analysis of Human Y-chromosome Haplotypes. American Journal of Human Genetics, 70,1197-1214.

49. Coia, V. , Destro-Bisol,G., Verginelli F., Battaggia,C., Boschi,I.,, Cruciani,F.,Spedini,G., Comas,D., and Calafell,F. ( 2005) Brief communication: mtDNA variation in North Cameroon: lack of Asian lineages and implications for back migration from Asia to sub-Saharan Africa, American Journal of Physical Anthropology ,(electronically published May 13, 2005; accessed August 5, 2005).

(http://www3.interscience.wiley.com/cgi-bin/fulltext/110495269/PDFSTART

50. Henn BM, Gignoux CR, Jobin M, Granka JM, Macpherson JM, Kidd JM, Rodríguez-Botigué L, Ramachandran S, Hon L, Brisbin A, Lin AA, Underhill PA, Comas D, Kidd KK, Norman PJ, Parham P, Bustamante CD, Mountain JL, Feldman MW. Hunter-gatherer genomic diversity suggests a southern African origin for modern humans. Proceedings of the National Academy of Sciences US A. 2011 Mar 29;108(13):5154-62. Epub 2011 Mar 7.
http://www.pnas.org/content/108/13/5154.full

51. Wood,E.T., Stover,D.A., Ehret,C., Destro-Bisol,G., Spedini,G., McLeod, H., Louie,L., Bamshad,M., Strassmann,B.I., Soodyall,H., Hammer,M.F. (2005) Contrasting patterns of Y-chromosome and mtDNA variation in Africa:evidence for sex-biased demographic processes. European Journal of Human Genetic, 13,867-876.

52. Berniell-Lee, G., Calafell, F., Bosch ,E. ,Heyer, E, Sica, L., Mouguiama-Daouda,| P., van der Veen, L., Hombert, J-M., Quintana-Murci , L.and, Comas, D. (2009) Genetic and Demographic Implications of the Bantu Expansion: Insights from Human Paternal Lineages, Molecular Biology and Evolution. 26(7),1581-1589; doi:10.1093/molbev/msp069

53. Winters C. (2011b). Is Native American R Y-

Chromosome of African Origin?. Current Research Journal of Biological Sciences 3(6): 555-558, 2011

54. Gonzalez, et al (2012) The genetic landscape of Equatorial Guinea and the origin and migration routes of the Y chromosome haplogroup R-V88. European Journal of Human Genetics advance online publication 15 August 2012; doi: 10.1038/ejhg.2012.167

# Chapter Six: The Paleoamericans came from Africa

## Abstract

The Paleoamericans are classified phenotypically as African, Australian or Melanesian based on multivariate methods and quantitative analysis . This grouping should only be Sub-Saharan African and Australian populations because the Melanesians and Sub-Saharan Africans share the same craniometric measurements. The craniometrics illustrate that PaleoIndians belonged to the Black Variety , but they do not allow us to establish conclusively where the Paleoamericans originated.

Some researchers believe the Paleoamericans came from East Asia across the Beringa Straits or from Europe because of the Solutrean tools found throughout North America . These points of origination are unlikely because the Ice shelf in the Northern Latitudes would have prevented passage from these destinations to South America where the oldest Paleoamerican sites have been excavated. The most likely place the Paleoamericans came from was Africa which is closer to the Americas, than either Europe or East Asia, and also the location where the Solutrean culture originated , and later expanded into Iberia.

## Introduction

What population is represented by the Paleoamericans skeletons ? . Some researchers claim the first Native Americans were mongoloid people who crossed the Baringa straits to enter the American continent (Hrdlička, 1907,1912), while other researchers claim they belonged to a different race ( Neves and Puciarelli, 1989, 1990, 1991).

Controversy surrounds the origin of the paleoamericans. Hrdlička (1907,1912) advanced the idea that the Paleoamericans were homogenous, a people that originated in East Asia or Melanesia. Other researchers were not so sure.

Dixon (2001) Imbelloni (1938) and Rivet (1908,1943), did not see the paleoamericans as a unitary population from East Asia, they felt that this population was probably more diverse. Even though there was some debate on the origin of the Paleoamericans Hrdlička's (1907,1912) ideas prevailed and researchers began to accept the idea East Asia was the homeland of the Paleoamericans.

In the 1960's there was a return to the study of craniometric quantitative analysis and multivariate methods to determine the Native American population (Neves , Powell and Ozolins, 1998,1999a,199b; Powell, 2005). This research indicated that the ancient Americans represent two populations, paleoamericans who were phenotypically African, Australian or Melanesian and a mongoloid population that appears to have arrived in the Americas after 6000 BC. Although we are sure of the ethnic identity of the paleoamericans we do not know from which continent the Paleoamericans came from.

Most of the earliest Paleoamerican sites dating between 65-13kya are found along the eastern coastline bordering on the Atlantic Ocean. This suggest that the first Americans probably came to the New World from Africa by boat not across the Beringa which was covered with ice long after the first Americans were living in South America (Imhotep, 2011). In this paper we will attempt to identify where the Paleoamericans originated.

**Method and Materials**

This is a review article. The author examined the database relating to the skeletal and cranial morphology of the PaleoIndians, using W.W. Howell's measurements these researchers determined the PaleoIndians were of African, Australian or Melanesian origin.

In addition to the skeletal evidence we looked at the archaeological databases of Eurasia and Africa to determine the probable origin for the Paleoamericans. The craniometric, anthropological and archaeological evidence was compared to the skeletal evidence, environmental factors and nautical histories to infer the probable continent of origin for the Paleoamericans.

## Results

We have good evidence concerning the ethnic identity of the Paleoamericans dating to 12kya. Archaeologists have excavated many sites in the Americas where they have recovered the skeletal remains of the Paleoamericans.

In the 1970's in Brazil an interesting skull of a girl was found. This skull was reconstructed and dated back to 12,000 BP (Neves and Pucciarelli, 1991; Neves, Powell and Ozolins, 1999c, 1999d). Dr. Walter Neves professor of biological anthropology at the University of Sao Paolo, after reconstructing the "Luzia" skull found that this personage was either an African or Pacific island type Black (Neves , Powell and Ozolins, 1999c ) .

Scientists have used the skulls of these skeletons to reconstruct the face of the Paleoamericans. The skulls of these Paleoamericans are of Native American females. The scientists gave them names Penon woman, Luzia and Naia.

The Paleoamericans are ethnically different from contemporary Native Americans. All of the Paleoamericans have been classified as part of the Black Variety. This includes Naia, and Penon Woman of Mexico and Luzia of Brazil ( See: Figure 2).

The craniometric mesasurements of the Paleoamerican skeletons fall within the Black Variety of homo sapien sapiens: African, Australian and the Melanesian phenotypic range (Neves, Powell and Ozolins,1998, 1999a,1999b; Powell,2005). The craniometric measurements of the PaleoIndians match the multivariate standard deviations of these three populations.

The determination of the Paleoamericans as members of the Black Variety is not a new phenomena. Howells ( 1973,1989,1995) using multivariate analyses, determined that the Easter Island population was characterized as Australo-Melanesian, while other skeletons from South America were found to be related to Africans and Australians ( Coon, 1962; Dixon, 2001;

Howell, 1989, 1995; Lahr, 1996). The African-Australo-Melanesian morphology was widespread in North and South America. For example skeletal remains belonging to the Black Variety have been found in Brazil (Neves, Powell, Prous and Ozolins,1998; Neves, Powell, Ozolins, 1998), Columbian Highlands (Neves, Pacciarelli, Munford, 1995; Powell, 2005 ), Mexico ( Gonza'lez-Jose, 2012), Florida ( Howells, 1995), and Southern Patazonia ( Neves, Powell and Ozolins,1999a,1999b).

In Figure 2, we have the reconstructions of Paleoamericans and the first European. The facial reconstruction of the Paleoamericans were startling ( Neto and Santo,2010). The bioanthropologist Walter Neves's reconstruction evidenced Negroid features for the Paleoamerican we call Luzia . Negroid features common to the Black Variety that were different from the indigenous mongoloid features of contemporary Americans ( Neto and Santo,

2010). What made this finding startling was that Neves using the mahalanobis distance and principal component analysis, found that 75 other skulls from Lagos Santa, were also phenotypically African or Australian ( Neves, Gonza'lez-Jose, Hubbe, Kipnis et al,2004). This has led researchers to highlight the fact that the PaleoIndians non-Mongoloid morphology was widespread across the Americas and that the population type is African-Australian (Munford et al, 1995; Neves et al, 2004; Neves and Hubbe, 2005). As a result, the cranial morphology of the ancient Americans indicates that two populations settled the Americas one African-Australian and the other mongoloid ( Neves and Hubbe,2005; Powell, 2005).

There is no single phenotypical negro that can be classified as Sub-Saharan African, so we have to apply the term Black Variety to the African-Australian-Polynesian populations.

Several types of blacks or negroes entered the Americas including the Anu or negrito type, Khoisan type, Australian and the Proto-Saharan or modern Sub-Saharan African black variety .

There is no single type of Negro or Black person. As a result, there are craniometric difference between Australoids /Australians , Mongoloids and Melanoids/ Sub-Saharan Africans (Laubenfels, 1968); craniometric differences that indicate at least two migrations of the Black Variety into Paleolithic Eurasia. Tsuenehiko Hanihare discussed the phenotypic variations between these populations ( Hanihare, 2005)..

Tsuenehiko classified these people into three major populations Southeast Asian Mongoloids (Polynesians), the Australians or Austroloid type and the Nicobar and Andaman (Melanoid/Sub-Saharan African type) samples which he found lie between the predominately Southeast Asian and Australoid/Australian type (Laubenfels, 1968; Hanihare, 2005). Sub-Saharan Africans and Melanesians share the same multivariate measurements (Winters, 2014b).

D.J de Laubenfels (1968) discussed the variety of Blacks found in Asia . The Australian aborigines and Melanesians show cranonical variates and represent two distinct Black populations(Laubenfels, 1968 ). The Australoids or Australians live mainly in Australia and the highland regions of Oceania, the Melanoid people on the otherhand live in the coastal regions of Near Oceania and Fiji (Winters, 2014b).

The Australian aborigines and Melanesians show cranial variates and represent two distinct Black populations (Laubenfels, 1968). The Australoids or Australians live mainly in Australia and the highland regions of Oceania, the Melanoid people on the otherhand live in the coastal regions of Near Oceania and Fiji ( Winters, 2014b) . The Melanoid people are recent migrants from Africa and mainland East Asia (Winters, 2014).

Other differences between these Black populations include Negroid / Melanoid brows being vertical and without eyebrow ridges, whereas Australoid brows are sloping and with prominent ridges (Laubenfels, 1968).

## Discussion

There are two scenarios propagated for the origin of the Paleoamericans . The first theory is that the Paleoamericans crossed from East Asia along the Beringa Straits or sailed to the Americas from East Asia. The second hypothesis, is that Paleoamericans entered the Americas from Europe due to the presence of Solutrean blade tools found in the Americas.

It is obvious that there were Paleo-americans that had either African or Australian features (Coon, 1962; Howells, 1973,1989, 1995; Lahr,1986;Powell, 2005). This suggest two migrations of Blacks into the Americas. One between 100- 50kya and another migration 20-13kya.

The first people to enter the Americas may have been the Australian type. The Australians did not leave Africa to settle much of Eurasia until probably 65kya, as supported by ancient sites in India that correspond to sites in Southern Africa.

A migration from Europe and or East Asia seems highly unlikely 20-30kya because of the Ice Age which would have made travel along the edge of the Atlantic and Pacific Ocean Arctic ice sheet too difficult (See Figure 1) . There was nothing in the Atlantic Ocean between Africa and the Americas to hinder sea travel.

Neves et al argues that the Paleoamericans came from East Asia because of the amh remains found at the Zhoukoudian Cave. Weidenreich (1939) found hominid fossils in the Upper and Lower Cave at Zhoukoudian. The individual in the Lower Cave was a Homo Erectus hominid (Sinanthropus pekinensis ) , and in the Upper Cave he found Oceanic or Melanesoid skeletons (Chang, 1977; Weidenreich, 1939). The Melanesoid skeletons are dated between 24-27kya (Sanz, 2014).

There are two major problems with the East Asia theory. First, the Ice shelf was too thick to make an overland trek into North America 27kya (See: Figure 1). Secondly, the Melanesoid people do not expand out of China until the expansion of the Lapita culture onto the Pacific Islands between 1600-500 BC (Winters,2014b).

The Paleoamericans were in South America at least between 65-48kya (Guidon and

Arnoud,1991; Guidon and Delibris, 1986, Guidon et al,1996; NYT,2015). This placed Paleoamericans almost 20,000 year in South America before they appear in East Asia. The archaeological evidence and Ice shelf in East Asia forces us to reject the Neves hypothesis.

The oldest North American culture is the Colvis culture. There is no archaeological evidence that situate the Clovis people in Siberia (Stanford and Bradley,2012).

Stanford and Bradley (2012) maintain that sites dating between 25,000-13000 years ago, namely the offshore Cinmar site, Meadowcroft Rock Shelter in Pennsylvania, Oyster Cove on the Chesapeake Bay, Cactus Hill in Virginia, and the Miles Point site have tool kits not found in Siberia. They claim that tools at these site resemble Solutrean tools, not Eurasian tool kits (Stanford and Bradley ,2012). "The majority of the oldest dated sites in the Americas with undisputed

artifacts are in the Chesapeake Bay region,"
wrote Stanford and Bradley (2008); "The
artifacts from these LGM sites are
technological and functional equivalents of
artifacts from the same period found in
southwestern Europe and are not
technologically or morphologically related
to any East Asian technology" (p.246).

The proposed Solutrean European
migration route was unlikely . Westley and
Dix (2008) illustrate the European
migratory route to America was highly
unlikely, and the data indicates that the
corridor probably did not exist.

 Sailors from Europe attempting to follow
the coastline from Europe to Canada
between 26-13kya would have had to brave
glaziers and Ice Age temperatures far below
zero. This would have made it impossible to
reach North America safely directly from
Europe (Westley and Dix, 2008).

Instead of the paleoamericans migrating from Eurasia, they probably made their way to the Americas directly from Africa ( Imhotep,2011) . The voyage from Africa-- is a shorter distance to the Americas than Europe. In addition, paleoamerican sailors could have made their way to the Americas on Currents, especially the Gulf Stream, that regularly flow from Africa, to the Americas.

Paleoamerican sites date between 65-10kya (Guidon and Arnoud,1991; Guidon and Delibris, 1986, Guidon et al,1996; NYT,2015; Winters,2014 ). This suggest that paleoamericans probably made several migrations from Africa. The first paleoamericans settled South America between 65-25kya (Guidon and Arnoud,1991; Guidon and Delibris, 1986, Guidon et al,1996; NYT,2015 ). A second paleoamerican migration took amh into North America, Brazil and Mexico 22-10kya.

This second migration would have included the ancestors of Luzia and Naia.

Today archaeologists have found sites from Canada to Chile that range between 20,000 and 65,000 years old (Imhotep, 2011; Guidon and Arnoud,1991; Guidon and Delibris, 1986, Guidon et al,1996; NYT,2015 ). There are numerous sites in North and South America which are over 35,000 years old.  These sites are the Old Crow Basin (c.38,000 B.C.) in Canada; Orogrande Cave (c.36,000 B.C.) in the United States; and Pedra Furada (c.45,000 B.C.) (Imhotep,2011).  Given the fact that the earliest dates for habitation of the American continent occur below Canada in South America is highly suggestive of the fact that the earliest settlers on the American continents came from Africa before the Ice melted at the Bering Strait and moved northward as the ice melted ( Bray,1988; Man's New World,1991; Haynes,1988).

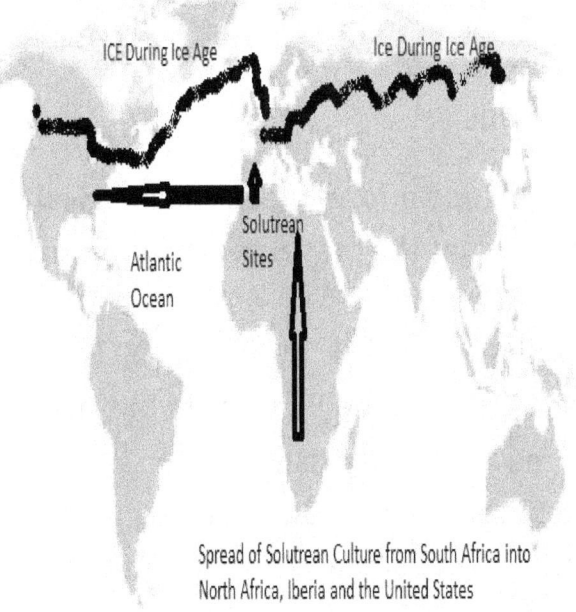

Spread of Solutrean Culture from South Africa into
North Africa, Iberia and the United States

## Figure 1: Spread of Salutrean Culture from Africa to North America

Dr. Guidon who conducted excavation at
the Pedra Furada site has found evidence of
human occupation dating back between
48,000-100,000 years old (Guidon and

Arnoud,1991; Guidon and Delibris, 1986, Guidon et al,1996; NYT,2015) . She proved that the tools are the result of human craftsmanship .

It would appear from the archaeological evidence that the first anatomically modern humans had made their way to Brazil 100kya ( NYT, 2015). This is 35,000 years before amh entered Eurasia. At this early date the Ice shield was too extensive for amh to have sailed from East Asia to the America, since amh did not enter Eurasia until 65kya based on recent models for the Out of Africa (OoA) event. And it was not until 27kya amh were established in China ( Sanz,2014).

It is becoming clear that people may have left Africa 100kya, instead of 60kya to settle the rest of the world. This may indicate that Proto-Australians in Africa made their way to America before the Khoisan since they represent probably the first amh to exit Africa.

Dr.Nieda Guidon hypothesized that man appeared in Brazil 100,000 years ago from Africa (NYT, 2015). She illustrated that her hypothesis was confirmed by 1) structures to make fire, i.e. hearths,2) stone tools and charcoal was found in the hearths that date back 100kya, 3) the Ice Age prevented people from reaching Brazil from Asia, while the winds and currents would have carried people directly from Africa to Brazil (NYT, 2015).

The charcoal and tools at Pedra Furada were found in hearths, sites of proposed human habitation. If the charcoal and tools were made naturally the entire site would have been burned, instead of just artifacts found in the hearths. We can accept Dr.Nieda Guidon hypothesis because it is normal science to use charcoal recovered from hearths to date a human habitation site (Guidon and Arnoud,1991; Guidon and Delibris, 1986, Guidon et al,1996; NYT,2015 ) .

Fire unless the result of lightening is produced by man. The evidence that fire existed in Brazil 65kya is an indication that man was at the site 65,000 years ago, since researchers found charcoal, which is the result of fire making (NYT, 2015).

The Khoisan were probably the ancestors of the paleoamericans who reached South America 48kya( Winters,2015a, 2015b).The question remains why did Africans 48kya discover South America (Weber,2015). The best answer is the spirit of adventure and discovery. At this time Africa was more wetter and the frequency of boat engravings in the Sahara indicate Africans had a high boat technology and navigation ability.

Around 100kya there were numerous lakes, rivers and streams in Africa that exited in the Atlantic Ocean. The distance from Lake Chad to Lake Congo was greater than the distance from Africa to Brazil. Any captain and sailors who had traded with cities and towns situated on Lake Mega Chad would have been familiar with storing enough foods to last long voyages.

The people around Lake Mega Chad had boats 15-8kya. Archaeologist excavated the Dafuna boat (Breunig, 1996). The Dafuna boat was found in Nigeria, near the Komodugu Gena River, and centered around Lake Mega Chad. This boat is evidence Africans probably knew navigation and sailed great distances around this Lake, that had rivers and streams that emptied in the Atlantic Ocean.

These ancient African navigators were probably like Columbus. They may have not known about South America, but they were willing to take a chance to see what lands lay at the edge of the Sea.

The Khoisan migrated across Africa over 50 kya from South Africa. The Khoisan carry the LOd haplogroup and L3. Haplogroup LOd is found at the root of human mtDNA. The TMRCA for LOd is 106kya (Gondor et al,2006). This makes haplotype AF-24 much older than L3a and probably explains why this haplotype is found among the Khoisan ( Chen, 2000).

The most archaic AMH remains come from Florished, South Africa; they date between 190-330 kya. Other ancient fossil evidence of AMH in South Africa come from Broken Hill (c.110kya) and the Klasis River caves (c. 65-105kya).

The Khoisan early migrated into North Africa. As a result, we see shared cultural and behavioral traditions between 200-40kya among South Africans and Moroccans.

The Neanderthal used Mousterian tools. These tools were also being used in Africa as early 130kya. This places Neanderthalers in North Africa.

The human types associated with the Neanderthal tools found at Jebel Ighoud and Haua Fteah resemble contemporaneous European Neanderthaler tools. The presence of Mousterian tools suggest that Neanderthalers mixed with Africans because we know that anatomically modern humans were living in the area at the time.

The African Neanderthal people used the common Levoiso-Mousterian tool kit originally discovered in Europe. The Nenderthal skeletons have come from Djebel Irhoud and El Guettar in Morocco (Ki-Zerbo,1981). Later Neanderthal people used the Aterian tool kit. It was probably in Morocco that Neanderthal and Khoisan interacted.

South African Khoi and San (SAK) dominated North Africa before other African populations and the Vandals migrated into North Africa. This is supported by Berber oral traditions.

In a Summary of three chapters dedicated to Africa, taken from the books **The Living Races of Man** (1965), Anthropology A to Z (1963) and **The Races of Europe** (1939), by Carleton S. Coon, http://slavanthro.mybb3.ru/viewtopic.php?t=1051

Coon observed:

"Legends persist along the fringes of the Sahara about the presence of an earlier, non-Europid people. According to the paramount chief of the Ait Atta, when their ancestors first came down from the mountains to their present winter quarters in the Dades Valley they found that region occupied by yellow-skinned people whom they conquered and reduced to the status of agricultural serfs. Later these yellow people mixed with Negro slaves, producing the present-day serfs, who are called Haratin. Many of the Haratin resemble Hottentots.

In the Fezzan in southern Libya live a people - the so-called Duwwud or Dawwada (worm-folk) - who speak Arabic, hunt jerboas, raise a few dates, and above all harvest the salt lakes, where they live, for Artemesia, a brine shrimp that multiplies in prodigious numbers. These shrimp are dried and compressed into cakes, which the Duwwud trade to Arab caravans. The Duwwud also look like Hottentots. Other partly Bushman and partly Negroid people are also to be found in the Sahara."

We can clearly see from this excerpt that relic Hottentot, Khoi and San populations persisted in North Africa and the Sahara up until the present.

The existence of the most ancient haplogroups in the Atlas Mountains and among the Khoisan supports the view that there has been an influx of non Khoisan people into the area in the past 20ky, but, relic Khoisan population elements remain constant in the Atlas Mountains up until today, just like in East Africa. Coon maintains that the Haritan also include relic SAK population elements.

An exception to this norm are the Khoisan who share a phylogenic relationship with Altai Neanderthals (Prufer, et al, 2013). Many researchers claim that Africans have no relationship to the Neanderthals. But Prufer et al (2013) found that the Khoisan share more alleles with Altaic Neanderthal than Denisova.

In the Supplemental section of Prufer et al (2013) there is considerable discussion of the relationship between Neanderthals and Khoisan. In relation to the Altaic Neanderthal the non-Africans have a lower divergence rate than Africans between 10-20%. Prufer et al (2013) note little statistical difference between non-African and African divergence.

Researchers have observed a relationship between the Neanderthals, the Khoisan and Yoruba. Prufer et al (2013) detected a relationship between the Neanderthal and Mandekan. It is interesting to note that Yoruba traditions place them in Mande-speaking areas (Prufer et al,2013).

There is interesting information in Prufer et al (2013) Figure S7.1. In Figure S7.1

the maximum likelihood tree of bonobo, Denisova and Neanderthal, the closest present-day humans are Africans, not Europeans (Prufer et al, 2013). Reading the Tree Chart Graph, the neighbor joining tree of archaic and present day human individuals has the Khoisan following the Denisova.

An interesting finding of Prufer et al (2013) was that Altaic Neanderthal and Denisova are estimated to have similar split times. The divergence estimate for African Khoisan-Mandekan and Altaic is younger than the split between Africans and Denisova archaic individuals and modern African individuals. The split times between the Khoisan and Mandekan may be explained by the presence of AF-24 haplotype in West Africa.

The Khoisan probably spread L3(M, N) into North and West Africa (Winters, 2010). In West Africa L3(M,N) is associated with the Senegambians haplotype AF24 (DQ112852), which is delineated by a DdeI site at 10394 and AluI site of np 10397 ( Gondor et al, 2006 ). The AF-24 haplotype is a branch of the African subhaplogroup L3 ( Chen, 2000). This is the same delineation of haplogroup M*.


The Khoisan carry haplogroups L3(M,N). Before they reached Iberia, the Khoisan probably stopped in West Africa on the way to North Africa.

Granted L3 and L2 are not as old as LOd, but Gonder et al (2006)provides very early dates for this mtDNA e.g., L3(M,N) 94.3; the South African Khoisan (SAK) carry L1c, L1,L2,L3(M,N) which date back to 142.3kya; the Hadza are L2a, L2, L3(M,N), dates to 96.7kya.

The dates for L1,L2,L3, M,N are old enough for the Khoisan to have taken N to West Africa, where we find L3, L2 and LOd and thence to Iberia as suggested in an earlier paper (Winters,2011).

It is interesting to note that LO haplogroups are primarily found among Khoisan and West Africans. This shows that at some point in prehistory the Khoisan had migrated into West Africa on their way to Morocco.

The basal L3(M) motif in West Africa is characterized by the Ddel site np 10394 and Alul site np 10397 associated with AF-24. This supports my contention that Khoisan speakers early settled West Africa on their way to Iberia.

The Khoisan may have introduced the L haplogroup to Iberia. The SAK populations carry haplogroups L2, and L3. Dominguez

(2005) ,noted that much of the ancient mtDNA found in Iberia has no relationship to the people presently living in Iberia today and correspond to African mtDNA haplogroups.

The SAK carry haplogroups L1c, L1,L2,L3 M,N and dates to 142.3kya; the Hadza are L2a, L2, L3, M,N, and dates to 96.7kya.

The dates for L1,L2,L3(M,N) are old enough for the Khoisan to have taken N to West Africa and thence Iberia.

Dominguez (2005) found that the lineages recovered from ancient Iberian skeletons are the African lineages L1b,L2 and L3. Almost 50% of the lineages from the Abauntz Chalcolithic deposits and Tres Montes, in Navarre are the Sub-Saharan lineages L1b, L2 and L3. The appearance of phylogenetically related sequences of hg L3 present in many ancient Iberian skeletons suggest that this haplogroup may have a long history in Iberia. This would support the possibility that SAK populations early settled ancient Iberia.

Anatomically modern humans arrived in Senegal during the Sangoan period. Sangoan artifacts spread from East Africa to West Africa between 100-80kya. In Senegal Sangoan material has been found near Cap Manuel (44), Gambia River in Senegal ( Davies, 1967; Wai Ogusu, 1973); and Cap Vert ( Phillipson, 2005).

The TMRCA of LOd dates to 106kya. As a result, anatomically modern humans (amh) had plenty of time to spread this haplogroup to Senegal. In West Africa the presence of amh date to the Upper Palaeolithic (Giresse, 2008). The archaeological evidence makes it clear that amh had ample opportunity to spread LOd and L3(M,N) which has an affinity to AF-24 (Chen et al, 2000), to West Africa during this early period of demic diffusion (Winters, 2010).

The earliest evidence of human activity in West Africa is typified by the Sangoan industry ( Pillipson ,2005). The amh associated with the Sangoan culture may have deposited Hg LOd and haplotype AF-24 in Senegal thousands of years before the exit of amh from Africa. This is because it was not until 65kya that the TMRCA of non-African L3(M,N) exited Africa ( Chang,1977) . Sangoan people may represent the earliest African population in Brazil that was 100-65,000 years old (Guidon and Arnoud,1991; Guidon and Delibris, 1986, Guidon et al,1996).

The Black Variety who represented Naia, and Luzia were probably the Khoisan people. The Khoisan 47kya had already settled in Europe. In Europe the Khoisan represents the Cro-Magnon people (Winters, 2008, 2011,2015).

The Khoisan were the Cro-Magnon people of Europe (Winters, 2008, 2011, 2014). They were the first anatomically modern humans to enter western Eurasia (Winters, 2011). The Khoisan probably introduced haplogroup M to western Eurasia (Winters, 2010,2011, 2014).

The Khoisan carry haplogroups L3(M, N). Before they crossed the Straits of Gibraltar to reach Iberia, they probably stopped in West Africa. The basal L3(M) motif in West Africa is characterized by the Ddel site np 10,394 and Alul site np 10,397 associated with AF-24 (Winters,2010). This supports my contention that Khoisan speakers early settled North and West Africa on their way to Iberia (Winters 2008).

Many North American Native Americans carry the X2 haplogroup. The American X Haplogroup is X2a . The Americas X2a is closely related to haplogroup X2j which is found among Egyptians (Fernandes et al. 2012). North African X2j shares two mutations with X2a at np sites 16,179 and

16,357. Fernandes et al. (2012), has suggested that the most likely place for the common ancestor of the American and African X2 populations was in North Africa. These researchers date the TMRCA of the X2 lived around 21kya in North Africa.

The dating of the TMRCA of X2, in North Africa 21kya corresponds to the dating for Solutrean culture in North Africa. This suggest that Paleoamericans introduced Hg X2 into North America.

It appears that the first Europeans were Khoisan (Boule and Vallois , 1957). They entered Western Europe across the Straits of Gibraltar ( Winters, 2008,2011). These people were Khoisan (Boule and Vallois , 1957). The Khoisan took their art and culture to Europe 40kya Boule and Vallois (1957). Here they constructed the Aurignacian, Grimaldi and Solutrean cultures (Boule and Vallois 1957; Winters, 2008,2011). Since the first Europeans had come from North Africa, we also find the Solutrean culture in Africa.

Many researchers have recognized that the Solutrean culture of Iberia probably originated in Africa ( Burkitt, 2012; Childe, 2001; Debenath et al, 1986; Debenath and Dibble,1994; Tiffagom, 2007 ) . It is the mainstream view of Spanish prehistorians that the Solutrean culture originated in Africa (Pericot, 1950). Boule and Vallois (1957) noted that ancient tool kits found in South African burials along the coast are associated with the Solutrean industry .


Pericot (1950,1955) believed that the tanged points at the Parpallo site of the Solutrean were of Aterian cultural origin. Burkitt ( 2012) said that there were Algerian tools similar to the Solutrean tool kit. Gordon Childe (2009) claimed that the North African and Spanish populations that used the Solutrean tools were in direct communication. By the 1960's, though, Smith (54) was able to reject the hypothesis of an African origin for the Solutrean culture.

The African hypothesis for the origin of the Solutrean culture has been revised by Debénath et al (1986) and Ramos (1998). Debénath et al ( 1986) argues that Iberomarusians crossed the Straits of Sicily, into Tunesia 25-22kya, and progressively drove the Solutreans out of North Africa into Iberia. Debénath et al (1986) maintains that this migration   OoA matches the origination of the Solutrean culture after 21kya. The Solutrean tanged points are at least 18-19ky old at Estremadura, Calderirao Cave and Parpalló Cave  in Valencia ( Straus, 2001).

Researchers have found evidence that Solutrean artifacts have been found on North American sites where PaleoAmericans remains have been found. The Solutrean people were Khoisan. This has led some researchers to create the so-called Solutrean hypothesis that proposes that ancient America was settled by ancient Europeans.

**Conclusion**

In summary , the tools found at the offshore Cinmar site, Meadowcroft Rock Shelter in Pennsylvania, Oyster Cove on the Chesapeake Bay, Cactus Hill in Virginia, and the Miles Point dating between 26-13 kya, appear to be similar to the Solutrean tools (Stanford and Bradley ,2012) . The Solutrean artifacts in the Americas probably relate to Khoisan who sailed from Africa to America.

The Solutrean culture originated Africa. North Africa is the location for the common ancestor of the American and African haplogroup X2 populations (Fernandes et al. 2012).

Given the short distance between   Africa
and Brazil, the first Paleoamericans
probably came directly to Brazil between
65-100kya from Africa , as evidenced by the
sites of human occupation found in Brazil
dating to this time (NYT, 2015). The fact
that the ancient people in Europe, Africa
and the Americas were phenotypically
Australian or Sub-Saharan African indicate
that for a considerable period of time the
world was dominated by populations with
dark skin belonging to the Black Variety
(Winters,2014a).

Although a migration from Europe seems
highly unlikely 20-30kya because of the Ice
Age. Ancient man could have made their
way to the Americas directly from Africa
which is a shorter distance to the Americas
than Europe. The rock art of Africa is rich in
boat engraving so we can infer that Africans
have long had the nautical ability to travel
by sea. Also ancient sailors could have
made their way to the Americas carried on

Currents, especially the Gulf Stream, that regularly flows from Africa, to the Americas.

In Figure 2, we see the ancient Americans and Europeans. Archaeologist have reconstructed the faces of ancient Americans from Brazil and Mexico. These faces are based on the skeletal remains dating back to 12,000BC.

Researchers agree that the first Americans, Naia of Mexico, Luzia of Brazil and Kennewick Man, found near the Columbia River in the State of Washington, were all phenotypically paleoamericans ( Neves and Pucciarelli, 1991; Neves, Powell and Ozolins, 1999c, 1999d ). This finding has added significance because the first Europeans were dark skinned and probably Khoisan ( Winters, 2014 ).

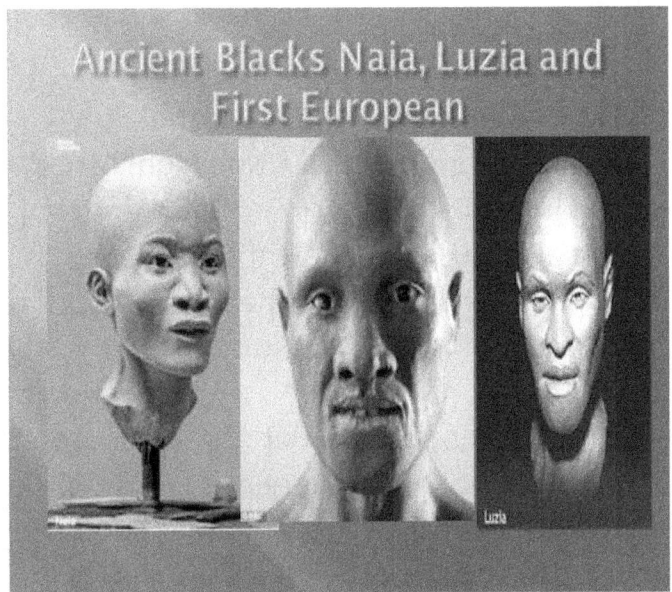

*Figure 2: Ancient Native Americans and the first European in the Center*

It appears that the first Europeans entered Western Europe across the Straits of Gibraltar. These people were Khoisan. The Khoisan took their art and culture to Europe 40kya ( Boule and Vallois ,1957;Winters, 2008,2011). Here they constructed the Aurignacian, Grimaldi and Solutrean cultures. Since the first Europeans had come from North Africa, we also find the Solutrean culture in Africa ( Boule and Vallois (1957; Burkett, 2012; Childe, 2009;

Ramos, 1998 ).

In Figure 1, we show the route the Paleoamericans probably took from Africa to the United states carrying Solutrean tool kits. It also illustrates how the Solutrean took kit originated in Southern Africa and was taken north by the Khoisan.

Africa is closer to the Americas than Europe. In the Atlantic Ocean there are Currents, that would have easily carried the Khoisan from Africa to the Americas. This view is supported by the fact that most ancient archaeological sites of paleoIndian habitation are nearer to the Atlantic Ocean, than the Pacific Ocean (Stanford and Bradley ,2012 ).

In addition, in Africa we find the Dafuna boat ( Breunig, 1996). The Dafuna boat has been dated to 8500 B.P., the culture associated with the people who built the Dafuna boat date back to 12,000 BP. This would indicate that around the time the paleoamericans: Kennewick man, Naia and

Luzia inhabited the Americas, Khoisan in Africa had the naval technology and nautical ability to have sailed to the Americas. Given the archaeological, and genetic evidence we can declare that the Paleoamericans came from Africa.

**References:**

Boule, M., HV Vallois . 1957. Fossil Man . Dryden Press New York

Breunig, P. 1996. The 8000-year-old dugout canoe from Dufuna (NE Nigeria), G. Pwiti and R. Soper (eds.), Aspects of African Archaeology. Papers from the 10th Congress of the PanAfrican Association for Prehistory and related Studies. University of Zimbabwe Publications (Harare ) 461-468.

Bray, W.(1988)."The Paleoindian debate". Nature 332, (10 March) 1988, p.107

Burkitt,M C. 2012. Prehistory: A Study of Early Cultures in Europe and the Mediterranean Basin.

Chang. K (1977). The archaeology of ancient China. New Haven, Yale University Press .

Chen YS, Olckers A, Schurr TG, Kogelnik AM, Huroponen K, Wallace DC. (2000). mtDNA variation in the South African Kung and Khwe— and Their genetic relationships to other African populations. Am J Hum Genet, 66(4): 1362-1383.

Childe, V. G. . (2009).The Prehistory of European Society.

Coon, C. S. 1962. The Origin of Races. New York: Knopf.

Davies,O. (1967). West Africa before the Europeans. London.

de Domínguez E.F. Polimorfismos de DNA mitocondrial en poblaciones antiguas de la cuenca mediterránea. Universitat de Barcelona. Departament Biologia Animal, 2005 (PhD thesis).

Débenath, A., Raynal, J.-P., Roche, J., Texier, J.P., Ferembach, D., 1986. Stratigraphie,habitat, typologie at devenir de l'Atérien Marocain: Données récentes. L'Anthropologie 90 (2), 233e246.

Débenath, A., Dibble, H., 1994. Handbook of Palaeolithic Typology. In: Lower and Middle

Palaeolithic of Europe, vol. 1. University of Pennsylvania Press.

Dixon, E. J. 2001. Human colonization of the Americas: timing, chronology and process. Quaternary Science Review, 20: 277–99.

Fernandes, V., F. Alshamali, M. Alves, M. D. Costa, J. B. Pereira, N. M. Silva, L. Cherni, et al. 2012. The Arabian Cradle: Mitochondrial Relicts of the First Steps along the Southern Route out of Africa.American Journal of Human Genetics 90: 347–55.doi:10.1016/j.ajhg.2011.12.010 .

Gonder MK, Mortensen HM, Reed FA, de Sousa A, Tishkoff SA.(2006).: Whole mtDNA Genome Sequence Analysis of Ancient African Lineages. Mol Biol Evol., Dec 28.

Gonza´lez-Jose´, R., Hernande´z, M., Neves, W. A., Pucciarelli, H. M. and Correal, G. 2002. Cra´neos del Pleistoceno tardio-Holoceno tempramo de Me´xico en relacio´n al patro´n morfolo´gico paleoamericano. Paper presented at the 7th Congress of the Latin American Association of Biological Anthropology, Mexico City.

Giresse,P. (2008). Tropical and sub-Tropical West Africa—marine and Continental changes during the late Quaternary. Volume 10. Elsevier Science.

Guidon, N. and Delibrias, G. 1986. "Carbon-14 dates point to man in the Americas 32,000 years ago." Nature 321:769-771.

Guidon, N., and B. Arnaud. 1991. "The chronology of the New World: Two faces of one reality." World Arch. 23(2):167-178.

Guidon, N., et al.1996. "Nature and Age of the Deposits in Pedra Furada, Brazil: Reply to Meltzer, Adovasio & Dillehay," Antiquity, 70:408.

Hanihare, T. (2005). Interpretation of craniofacial variations and diversification of East and Southeast Asia. In Bioarchaeology of Southeast Asia. (Eds.) Marc Oxenhan and Nancy Tayles (pp.91-111). Cambridge.

Haynes,Jr.,C V (1988)."Geofacts and Fanny". Natural History ,pp.4-12.

Howells, W. W. 1973. Cranial Variation in Man: A Study by Multivariate Analysis of Patterns of Difference among Recent Human Populations. Papers of the Peabody Museum of Archaeology and Ethnology, 67. Cambridge, MA: Harvard University.

Howells, W. W. 1989. Skull Shapes and the Map: Craniometric Analyses in the Dispersion of Modern Homo. Papers of the Peabody Museum of Archaeology and Ethnology,79. Cambridge, MA: Harvard University.Early Holocene human skeletal remains from Cerca Grande 497

Howells, W. W. 1995. Who's Who in Skulls: Ethnic Identification of Crania from Measurments. Papers of the Peabody Museum of Archaeology and Ethnology, 82. Cambridge. MA: Harvard University.

Hrdlička A. 1907. Skeletal remains suggesting or attributed to early man in North America. Bureau of American Ethnology Bulletin 33. Washington: Smithsonian Institution.

Hrdlička A. 1912. Early Man in South America. Bureau of American Ethnology, Bulletin 52. Washington: Smithsonian Institution.

Imbeloni J. 1938. Tabla classificatoria de los indios: regions biologicas y grupos raciales humanos de America. 12:229–249.

Imhotep, D. 2011. The First Americans Were Africans: Documented Evidence.

Kieran Westley and Justin Dix. (2008). The Solutrean Atlantic Hypothesis: A View from the Ocean Journal of the North Atlantic, 1:85-98.

Ki-Zerbo,J. (1981). Unesco General History of Africa Vol. 1: Methodology and African Prehistory .

Lahr, M. M. 1996. The Evolution of Modern Human Diversity: A Study of Cranial Variation.Cambridge: Cambridge University Press.

Laubenfels, D J. (1968). Australoids, Negroids and Negroes: A suggested explanation for their distinct distributions. Annals Association of Am. Geographers, 58(1), 1968: 42-50.

"Man's New World arrival Pushed back", Chicago Tribune, (9 May 1991) Sec.1A, p.40;and A.L. Bryan, "Points of Order". Natural History , (June 1987) pp.7-11.

Martin,P S. and R.G.Klein (eds.),Quarternary Extinctions: A Prehistoric Revolution, Tucson:University of Arizona Press,1989.

Munford, D., Zanini, M. C. and Neves, W. A. 1995. Human cranial variation in South America: implications for the settlement of the New World. Brazilian Journal of Genetics, 18: 673–88.

Neto, V V and Santo, R V.(2010). The color of the bones: Scientific Narrative and cultural appropriations of 'Luzia" a prehistoric skull from Brazil. Mana 5.

Neves, W. A. and Pucciarelli, H. M. 1989. Extra-continental biological relationships of early South American human remains: a multivariate analysis. Cieˆncia e Cultura, 41: 566–75

Neves, W. A. and Pucciarelli, H. M. 1990. The origins of the first Americans: an analysis based onthe cranial morphology of early South American human remains. American Journal of Physical Anthropology, 81: 247.

Neves, W. A. and Pucciarelli, H. M. 1991. Morphological affinities of the first Americans: an exploratory analysis based on early South American human remains. Journal of Human Evolution, 21: 261–73.

Neves, W. A. and Meyer, D. 1993. The contribution of the morphology of early South and

Northamerican skeletal remains to the understanding of the peopling of the Americas. American Journal of Physical Anthropology, 16 (Suppl): 150–1.

Neves, W. A., Powell, J. F., Prous, A. and Ozolins, E. G. 1998. Lapa Vermelha IV Hominid 1: morphologial affinities or the earliest known American. American Journal of Physical Anthropology, 26(Suppl): 169.

Neves, W. A., Powell, J. F. and Ozolins, E. G. 1999a. Extra-continental morphological affinities of Palli Aike, southern Chile. Intercieˆncia, 24: 258–63.

Neves, W. A., Powell, J. F. and Ozolins, E. G. 1999b. Modern human origins as seen from the peripheries. Journal of Human Evolution, 37: 129– 33.

Neves W.A . and Pucciarelli H.M. 1991. "Morphological Affinities of the First Americans: an exploratory analysis based on early South American human remains". Journal of Human Evolution 21:261-273.

Neves W.A ., Powell J.F. and Ozolins E.G. 1999. "Extra-continental morphological affinities of Lapa

Vermelha IV Hominid 1: A multivariate analysis with progressive numbers of variables. Homo 50:263-268

Neves W.A ., Powell J.F. and Ozolins E.G. 1999. "Extra-continental morphological affinities of Palli-Aike, Southern Chile". Interciencia 24:258-263.
http://www.interciencia.org/v24_04/neves.pdf

Neves, W.A., Gonza´ lez-Jose´ , R., Hubbe, M., Kipnis, R., Araujo, A.G.M., Blasi, O., 2004. Early Holocene Human Skeletal Remains form Cerca Grande, Lagoa Santa, Central Brazil, and the origins of the first Americans. World Archaeology 36, 479-501

Neves, W. A., and M. Hubbe. 2005. Cranial morphology of early Americans from Lagoa Santa, Brazil: Implications for the settlement of the New World. Proc. Natl. Acad. Sci. USA 102:18,309–18,314.

NYT (New York Times). (2015) Human's First Appearance in the Americas .
http://www.nytimes.com/2014/03/28/world/americas/discoveries-challenge-beliefs-on-humans-arrival-in-the-americas.html?hp&_r=4

Pericot, L., 1950. La Espan8 a Primitiva. Barcelona.

Pericot, L., 1955. Sur les connexions europeHennes de l'AteHrien. Etat actuel du proble`me. Actes du II Congre`s Panafricain de PreHhistoire (Algiers), Paris, p. 375.

Phillipson, D.W.(2005). African Archaeology. Cambrige.

Powell,J.F. (2005). First Americans:Races, Evolution and the Origin of Native Americans. Cambridge University Press.

Pruler,K, Racimo,F.,Patterson,N et al. (2014). The complete genome sequences of Neanderthal from the Altai, Mountains. Nature , 505/7481: 43-9. doi .10.1038/ Nature 12881.Epub.2013.Dec.18.

Ramos, J., 1998. La conexion norteafricana: panorama del Ateriense y su posible in#uencia en la conformacioHn del Solutrense en el sur peninsular. In: MartmHn, A., VelaHzquez, F., Bustamante, J. (Eds.), Estudios de la Universidad de CaHdiz Ofrecidos a la Memoria del Profesor Braulio Justel Calabozo. Universidad de CaHdiz, CaHdiz, pp. 437-445.

Rivet P. 1908. La race de Lagoa-Santa chez le populations precolombiennes de L'Equateur.

Bulletins et Memoires de la Societe D'Anthropologie de Paris 19:209–275.

Rivet, P. 1943. Les Origines de l'homme américain. Montreal: les Éditions de l'Arbre [reeditado en 1957 por Éditions Gallimard]

Sanz, Nuria . (2014). Human origin sites and the World Heritage Convention in Asia. UNESCO.

Scozzari, R, Massaia,A, Trombatta,B. et al.(2014). An unbiased resource of novel SNP markers provides a new chronology for human Y-chromosome and reveals a deep phylogenetic structure in Africa. Genome Research, January 6,2014, doi: 10.1101/gr./60785.113.

Smith, P.E.L., 1966. Le SolutreHen en France. Delmas, Bordeaux.

Stanford, Dennis J. and Bruce Bradley (2012). Across Atlantic Ice: The Origin of America's Clovis Culture. University of California Press.

Straus,L. G. (2001). Africa and Iberia in the Pleistocene, Quaternary International 75 (2001) 91-102

Tiffagom, M., 2006. El Solutrense de facies ibérica o la cuestión de los contactos mediterráneos

(Europa, África) en el Último Máximo Glacial. In:
Sanchidrián, J.L., Márquez, A.M., Fullola, J.M.
(Eds.), IV Simposio de Prehistoria Cueva de Nerja.
La Cuenca Mediterránea durante el Paleolítico
Superior 38000-10000 años. Reunión de la VIII
Comisión del Paleolítico Superior U.I.S.P.
Fundación Cueva de Nerja. Nerja, pp. 60e77.

Wai-Ogusu,A.(1973). Was there a Sangoan
industry in West Africa, West African Jour of
Arcaheo,3:191-96.

 Weber, G. (2015) Fuegian &PatagonianGenetics -
and the settling of the Americas. Retrieved on
3/20/2015 at: http://www.semiaquatic-
ancestors.nl/darwin_bronnen/fuegans_herkomst
_feiten_dna/text-FuegianGenetics.htm and
http://www.andaman.org/BOOK/chapter54/text-
Fuego/Genetics/text-FuegianGenetics.htm

Weidenreich. F. 1939. On the earliest
representative of modern mankind recovered on
the soil of East Asia( in BC.ull. Nat. Hist. Soc.
Peiping 13:161-173.

Winters, C. (2008). Aurignacian Culture: Evidence
of Western Exit for Anatomically Modern Humans.
South Asian Anthropologist, 8, 79-81.

Winters,C. (2010). The African Origin of mtDNA Haplogroup M1 . Current Research Journal of Biological Sciences 2(6): 380-389, 2010. https://www.academia.edu/3036833/The_Africa n_Origin_of_mtDNA_Haplogroup_M1

Winters, C. (2011). The Gibraltar out of Africa Exit for Anatomically Modern Humans. WebmedCentral BIOLOGY, 2. http://www.webmedcentral.com/article_view/23 11

Winters, C. (2014a). Were the First Europeans Pale or Dark Skinned? Advan in Anth, 4, 124-132. http://dx.doi.org/10.4236/aa.2014.43016

. Winters,C. (2014b). AFRICAN AND DRAVIDIAN ORIGINS OF THE MELANESIANS. Indian Journal of Fundamental and Applied Life Sciences , 4(3):694-704. http://www.cibtech.org/J-LIFE-SCIENCES/PUBLICATIONS/2014/Vol-4-No-3/JLS-103-JLS-073-JUN-CLYDE-AFRICAN-MELANESIANS.pdf

Winters, C. (2015a)African origins of Paleoamerican DNA, CIBTech Journal of Microbiology, 4 (1):13-18. http://www.cibtech.org/J-

Microbiology/PUBLICATIONS/2015/Vol-4-No-1/03-CJM-004-CLYDE-AFRICAN-DNA.pdf

**Winters C. 2015b. Inference of Ancient Black Mexican Tribes and DNA. WebmedCentral GENETICS ; 6(3)**
http://www.webmedcentral.com/article_view/4856

# Appendix One: Native American Slavery in the United States

The Black Native Americans were enslaved before the Transatlantic slave trade. During slavery Black and Mongoloid Native Americans mated frequently with West African slaves.

Before Europeans reached the United States the Native Americans held war captives. These individuals were captured during wars and held as captives.

The Native American war captives might work on the holdings of a tribe, be used in ritual sacrifice or exchanged by their captors when redeeming members of their own tribe. Over time the war captives were usually integrated in the tribes they were held captive.

As a result there was no trade in slaves.

Europeans introduced the idea of chattel slavery to the Native Americans. The first slaves in North America, beginning as early as 1620, were not Africans, they were Native Americans.

The French, British and Spanish found large tracts of land to cultivate when they arrived in the United States. There were few indentured servants or freemen to employ to work the land. As a result, Europeans acquired labor by forcing Native Americans to become their slaves. Europeans acquired slaves in a variety of ways, including, kidnapping, and staging wars against the Indians If the Indians lost the war their entire would be sold into slavery either in the United States or Caribbean.

The Black Native Americans (BNA) lived on valuable

farmlands during the Colonial period. The English

and Americans wanted this land. This led to violent conflicts between BNAs and white Americans. In

New England, the BNAs were eliminated by slaving,

warfare and forced removal.

The Europeans would see a small tribe and start a war

to take the Indian prisoners as slaves. The best example

of this was the Pequot War of 1636. The Europeans

practice a policy of divide and rule. They usually

solicited help from the Native Americans to capture

and destroy other tribes. In the Pequot War the

Narragansett, Mohegan and Niantic tribes helped

Colonists in Connecticut, Massachusetts, and Plymouth

to defeat the Pequot. All the Native Americans that

were not killed were sold into slavery for life. Seven

hundred of the Pequot were killed, and the women and

children became chattel slaves according to the court

records.

The French enslaved Native Americans around the

Great Lakes, Minnesota, Missouri Country and Lousiana

(Gallay, 2002). The Europeans also needed labor to

work the fields.

Originally the Europeans would purchase the captives

held by a Native American tribal groups. The Americans

used these Native Americans to work on tobacco

plantations in the South, and farms in the North.

# Pequot Indians

The Americans provided the Native Americans with

guns and cheap goods to purchase Native

American/Indian slaves. Between 1670 and 1720 many

BNAs were enslaved. The BNAs were sold into slavery

throughout the Thirteen Colonies, Canada and the

Bristish West Indies (Gallay, 2002). The majority of BNAs

sold into slavery, by white and Indian slave traders were the Choctaw (Gallay, 2002), and Yamasee and other Carolina tribes (Lauber, 1970; Newell, 2009).

The Black Native Americans did not accept slavery willingly. Because they knew the land they might often run away. As a result, Black Native American slaves were usually sold far away from their homes or in the Caribbean to prevent the Native Americans from running away.

In addition to forcing BNA into slavery to provide labor for the Colonists, Europeans also saw this as a way to acquire more land for themselves. As a result, after a war the entire tribe was usually sold into slavery or forced to move off their lands to another local.

The Native American slave trade decimated the Black Native American community. Between 1670 and 1715 , an estimated 40,000 to 50,000 BNA were sold into

slavery in the Caribbean. Most of these slaves were

Black Native Americans from the Carolinas.

References:

Gallay,A. (2002). The Indian Slave Trade: The Rise of the

English Empire in the American South, 1670–1717 .

Gilio-Whitaker D (2015). The Untold History of

American Indian Slavery, Retrieved 2/24/15

http://nativeamericanhistory.about.com/od/controvers

ies/a/The-Untold-History-Of-American-

IndianSlavery.htm

Most books on Native American and Afro-American

relations frame it within the context of Indian Master

and African slaves. These books include Gary Zellar,

African Creeks: Estelvste and the Creek Nation (2007),

James F. Brooks, Captives and Cousins: Slavery, Kinship, and Community in the Southwest Borderlands (2002); Tiya Miles, Ties That Bind: The Story of an Afro-Cherokee Family in Slavery and Freedom (2006); and Theda Purdue, Mixed Blood Indians: Racial Construction in the Early South (2005). .

# Appendix Two: Indigenous Indian DNA is admixed with African DNA.

The foundational mtDNA lineages for Mexican Indians are lineages A, B, C and D. The frequencies of these lineages vary among population groups. For example, whereas lineages A, B and C were present among Maya at Quintana Roo, Maya at Copan lacked lineages A and B (Gonzalez-Oliver, et al, 2001). Haplogroup A is found among Mixe and Mixtecs ( Bonilla et al, 2005). This supports Carolina Bonilla et al (2005) view that heterogeneity is a major characteristic of Mexican population.

The mtDNA A haplogroup common to Mexicans is also found among the Mande speaking people and some East Africans (Salas et al, 2002). The Mande speakers carry mtDNA haplogroup A, which is common among Mexicans (Jackson et al, 2005). In addition to the Mande speaking people of West Africa, Southeast Africa Africans also carry mtDNA haplogroup A ( Salas et al, 2002).

African y-chromosome are associated with YAP+ and 9bp. The YAP-⏾ associated with A-⏾G transition at

DYS271 is found among Native Americans. The YAP+ individuals include Mixe speakers (32-33). YAP+ is often present in haplogroups (hg) y-chromosome C and D.

# Appendix Three: Fuegians and Khoisan

Researchers believe the Fuegians are remnants of the earliest settlers of the New World. The Fuegians have different genetic make-up from the other South Americans Indians.

George Weber notes that: "As far as we can draw conclusions from a single skeleton, the fact that Pali Aike aligns with Africans and Australians, instead of with Asians and modern Amerindians is significant in at least two different ways for the current debate about who were the first Americans." .

## Researchers believe the Fuegians are remnants of the earliest settlers of the New World

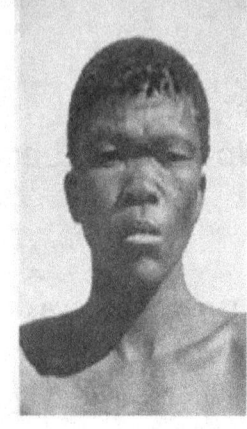

Khoisan

Fuegian

Figure 1: Fuegians and Khoisan

George Weber added that: "First, it shows that people similar to those that inhabited the Lagoa Santa area, in central Brazil, and the area of Sabana de Bogota, in Colombia, once had a wide distribution across South America, reaching even the southernmost region of the sub-continent." He added: Second, but intrinsically related to the first fact, that the non-Mongoloid morphology already demonstrated to occur in tropical and subtropical areas of South America.

The Fuegians and Khoisan share many culture traits including housing and tools. In Figures 5-6 , you can see the similar elements shared by the Fuegians and Khoisan.

Fuegians and wigwams

· Khoisan house

Figure 2: Fugian and Khoisan Dwellings

The Fuegians and Khoisan carry the M174 gene related to the D haplogroup. The Fuegians may carry the genes of the first Americans note the facial characteristics of the Fuegians and Khoisan and the homes they built.

Dr. Clyde Winters

www.ingramcontent.com/pod-product-compliance
Lightning Source LLC
Chambersburg PA
CBHW070313190526
45169CB00005B/1616